Biology for Nonbiologists

FRANK R. SPELLMAN

Government Institutes
An imprint of
THE SCARECROW PRESS, INC.
Lanham, Maryland • Toronto • Plymouth, UK
2007

 **Government Institutes**

Published in the United States of America
by Government Institutes, an imprint of The Scarecrow Press, Inc.
A wholly owned subsidiary of
The Rowman & Littlefield Publishing Group, Inc.
4501 Forbes Boulevard, Suite 200
Lanham, Maryland 20706
http://www.govinstpress.com/

Estover Road
Plymouth PL6 7PY
United Kingdom

British Library Cataloguing in Publication Information Available

Library of Congress Cataloging-in-Publication Data

Spellman, Frank R.
  Biology for nonbiologists / Frank R. Spellman.
     p. cm.
  Includes index.
  ISBN-13: 978-0-86587-421-3 (pbk. : alk. paper)
  ISBN-10: 0-86587-421-2 (pbk. : alk. paper)
  1. Biology. 2. Human biology. I. Title.
  QH307.2S65 2007
  570—dc22   2006100624

∞™ The paper used in this publication meets the minimum requirements of American National Standard for Information Sciences—Permanence of Paper for Printed Library Materials, ANSI/NISO Z39.48-1992.
Manufactured in the United States of America.

For Revonna Livingston Bieber
The Ultimate Biologist and Friend

DATE DUE

Contents

APPENDIX

Preface

Why a text on biology for nonbiologists? Answer: Based on personal experience, I have found that many professionals do have some background in biology, but many of these folks need to stay current. Moreover, many practitioners are specialists (e.g., epidemiologists, community health specialists, toxicologists, environmental science and occupational health and safety professionals, teachers, and so forth). Herein lays the problem—that is, specialization. My view is that practitioners (of any type) should be generalists with a wide range of knowledge, not just specialists whose span of knowledge may be narrow. Again, based on personal experience, my students who generalize their education—spreading out their exposure to include several disciplines in several aspects of environmental studies—are afforded many more opportunities to broaden their upward mobility compared to those who narrowly specialize. They have a much better chance to rise to upper management positions, where more prestige, responsibility, pecuniary rewards, and, yes, more headaches await them.

Biology for Nonbiologists provides an overview and review of the important aspects of biology in a way that can be easily understood, even if the user has not taken any formal biology courses. In this text, as with my previous science texts, I define the multidisciplinary *environmental practitioner* as anyone engaged in work directly related to environmental management, planning, impact assessment, environmental protection, or environmental compliance, including such activities as permitting, compliance auditing, regulatory review, research teaching, engineering, design, quality assurance, and implementation of environmental protection and control. Moreover, I include the occupational health and safety professional that designs, implements, and evaluates a comprehensive health and safety program to maintain and enhance health, improve safety, and increase productivity in my definition of *environmental practitioner*. Last, I loosely define (with some license) anyone with an active interest in environmental concerns as an *environmental practitioner*.

Along with basic biological principles, the text provides a clear, concise presentation of the consequences of biology's interaction with the environment. Although *Biology for Nonbiologists* is primarily designed for professionals (especially practicing environmental professionals) who need a quick, rapid review of biological principles and applications, it is also designed for a more general audience. Even if you are not tied to a lab bench, I provide you—the nonbiologist—with the jargon, concepts, and key concerns of biology and biological practice. This book is compiled in an accessible, user-friendly format, unique in that it explains scientific concepts with little reference to mathematics and/or physical science.

The text begins broadly with a review of chemistry, cells, cell division, laws, reactions, photosynthesis, genetics, plants, animals, and biological diversity. I focus on the fundamentals of environmental biology and ecology as they apply to environmental regulatory compliance programs. Next, the focus shifts to human biology. This is the case, of course, because in regard to scientific knowledge nothing is more important to humans than human biology. Later, we move to more advanced discussions of community and ecosystem dynamics.

This book is designed to serve the needs of students, teachers, consultants, and technical personnel in environmental compliance positions who must review biological basics and fundamentals. The text is packed with information that students need to get ready for SAT exams. In order to maximize the usefulness of the material contained in the text, I have written and presented in plain English, and in a simplified and concise format.

Each chapter ends with a chapter review test to help evaluate mastery of the concepts presented. Before going on to the next chapter, you should take the review test, compare your answers to the key provided in the appendix, and review the pertinent information for any problems you missed. If you miss many items, review the whole chapter.

✓   **Note:** The checkmark symbol displayed in various locations throughout this text emphasizes an important point or points to study carefully.

Again, this text is accessible to those who have no experience with biology. If you work through the text systematically, you will acquire an understanding of and skill in biological principles—adding a critical component to your professional knowledge.

Frank R. Spellman
Norfolk, VA

Part I

GENERAL BIOLOGICAL PRINCIPLES

CHAPTER 1

Introduction

If it's green or wriggles, it's biology.
If it stinks, it's chemistry.
If it doesn't work, it's physics.

—Handy Guide to Science

Topics in This Chapter

- Why Study Biology?
- What Is Life?
- Levels of Organization
- Major Theories of Biology
- The Scientific Method
- Text Themes

Welcome to biology, the study of life. We are living things. Biology helps us to understand how we function and how we fit into the scheme of life—into our environment. Biology is a branch of science. Study of biology is a way of knowing the natural world. Biology is integrated with other sciences. It is one of the critical components or elements of the natural world—organisms are subject to laws of physics and chemistry. Mathematics is used in biology to analyze and interpret biological data. In this text we combine biology, humans and the environment—to focus on environmental biology—because they are inseparable. The mantra, the theme woven throughout the fabric of this text, is quite simple but critical: "Humanity belongs to the planet rather than vice versa" (Quinn 1995). The following is a comprehensive definition of biology.

bi•ol•o•gy n. **1.** the science of life and of living organisms, including their structure, function, growth, origin, evolution, and distribution. It includes botany and zoology and all their subdivisions. **2.** the life processes or characteristic phenomena of a group or category of living organisms: the biology of bacteria. **3.** the plant and animal life of a specific area or region.

The first part of this definition taken from the *American Heritage Dictionary of the English Language* captures many of the essential ingredients of biology (although definitions 2 and 3 might make a more challenging, interesting, and entertaining text).

So, how would you define biology? Are you tempted to answer that, like math and chemistry, it is just a prerequisite for other subjects of study? Biology is that, but much more. Again, biology explains our living world around us. Biology is the science that helps us to understand nature. Biology is a systematic study. Biology is the study of the composition and properties of life. Biology is the science that helps to explain the human endeavor. Biology is the study of connections between the everyday world and the natural world. Biology is so complex that no one person could expect to master all aspects of such a vast field, so it has been convenient to divide biologists into specialty areas. These specialty areas, disciplines, span almost every letter in the English alphabet A through Z, as shown in Table 1.1.

From such a large discipline pool, you might correctly expect that there are several career paths open to those with biology training. For example, a few of the career paths include research, health care, environmental management and conservation, education—including at the university, primary and secondary levels—science museums, zoos, aquariums, parks and nature centers. In addition, there are many careers for biologists who want to combine their scientific training with interests in other fields such as biotechnology, forensic science, politics and policy, business and industry, economics, mathematics, and science writing, communications and art. A few of the specific career fields available to biologists include:

- Bacteriologist
- Pathologist
- Medical secretary
- Dermatologist
- Paramedic
- Physiotherapist
- Horticulturalist
- Agronomist
- Landscape gardener
- Landscape architect
- Geneticist
- Plant pathologist
- Botanist
- Zoologist
- Fish hatchery operator
- Game warden
- Park superintendent
- Animal attendant
- Veterinarian
- Biological technologist

- Molecular biologist
- Medical biophysicist
- Dietitian
- Food scientist
- Microbiologist
- Biologist
- Biochemist
- Teacher
- Biological technologist
- Biological oceanographer
- Biological oceanographer
- Medical scientific illustrator
- Paleontologist
- Anthropologist
- Dental hygienist
- Dentist
- Nursing assistant
- Nurse
- Lab technician
- Nuclear medical technician

You are beginning the study or review of one of the most interesting subjects that you will ever come across. Of course, what you get out of studying biology will depend

Table 1.1. Biology Disciplines A through Z

Discipline	Description
Aerobiology	Studies organic particles, such as bacteria, fungal spores, very small insects and pollen, which are passively transported by the air (Spieksma 1991). One of the main fields of aerobiology has traditionally been to measure and report quantities of pollen as a service to allergy sufferers (Larsson 1993).
Anatomy	The bodily structure of a plant or an animal or of any of its parts. The science of the shape and structure of organisms and their parts.*
Arachnology	Study of spiders or scorpions or relatives of spiders.
Astrobiology	The study of the origin, evolution, distribution, and destiny of life in the universe.
Bacteriology	Study of bacteria, especially in relation to medicine and agriculture.*
Biochemistry	The study of the chemical substances and vital processes occurring in living organisms; biological chemistry; physiological chemistry.*
Bionics	Application of biological principles to the study and design of engineering systems, especially electronic system.*
Biogeography	The study of the geographic distribution of organisms.*
Bioinformatics	"The application of computer technology to the management of biological information" (Bordenstein 2006).
Biomechanics	The study of the mechanics of a living body, especially of the forces exerted by muscles and gravity on the skeletal structure.*
Biophysics	The science that deals with the application of physics to biological processes and phenomena.*
Biotechnology	The use of microorganisms, such as bacteria or yeasts, or biological substances, such as enzymes, to perform specific industrial or manufacturing processes.*
Botany	The science or study of plants.*
Cell Biology	Study of the cells.
Chorology	The study of the causal relations between geographical phenomena occurring within a particular region.
Cladistics	A system of classification based on the phylogenetic relationships and evolutionary history of groups of organism.*
Cryptozoology	The study of creatures, such as the Sasquatch, whose existence has not been substantiated.*
Cytology	The branch of biology that deals with the formation, structure, and function of cells.*
Developmental Biology	The study of the process by which organisms grow and develop.
Ecology	The science of the relationships between organisms and their environments.*
Entomology	The scientific study of insects.*
Epidemiology	The branch of medicine that deals with the study of the causes, distribution, and control of disease in populations.*
Ethology	The scientific study of animal behavior, especially as it occurs in a natural environment.*
Endocrinology	The study of the glands and hormones of the body and their related disorders.*

Table 1.1. Biology Disciplines A through Z (*continued*)

Discipline	Description
Evolutionary Biology	The discipline that describes the history of life and investigates the processes that account for this history.
Felinology	The study of cats.
Freshwater Biology	Studies the life and ecosystems of freshwater bodies.
Genetics	Branch of biology that deals with heredity, especially the mechanisms of hereditary transmission and the variation of inherited characteristics among similar or related organisms.*
Genomics	The study of all of the nucleotide sequences, including structural genes, regulator sequences, and noncoding DNA segments, in the chromosomes of an organism.*
Herpetology	Branch of zoology that deals with reptiles and amphibians.*
Histology	The anatomical study of the microscopic structure of animal and plant tissues.*
Ichthyology	Branch of zoology that deals with the study of fishes.*
Immunology	Branch of biomedicine concerned with the structure and function of the immune system, innate and acquired immunity, the bodily destination of self from nonself, and lab techniques involving the interaction of antigens with specific antibodies.*
Lichenology	Branch of biology that deals with the study of lichens.*
Limnology	The scientific study of the life and phenomena of fresh water, especially lakes and ponds.*
Malacology	Branch of zoology that deals with mollusks.*
Mammalogy	Branch of zoology that deals with mammals.*
Marine Biology	Study of ocean plants and animals and their ecological relationships.
Microbiology	Branch of biology that deals with microorganisms and their effects on other living organisms.
Molecular Biology	Branch of biology that deals with the formation, structure, and function of macromolecules essential to life, such as nucleic acids and proteins, and especially with their role in cell replication and the transmission of genetic information.*
Morphology	Branch of biology that deals with the form and structure of organism without consideration of function.*
Mycology	Branch of botany that deals with fungi.*
Myrmecology	Branch of entomology that deals with ants.*
Nematology	Branch of zoology that deals with nematodes.*
Oncology	Branch of medicine that deals with tumors, including study of their development, diagnosis, treatment, and prevention.*
Paleontology	Study of the forms of life existing in prehistoric or geologic times, as represented by the fossils of plants, animals, and other organisms.*
Paleobotany	Branch of paleontology that deals with plant fossils and ancient vegetation.*
Paleozoology	Branch of paleontology dealing with the recovery and identification of animal remains from archaeological

	contexts, and their use in the reconstruction of past environments and economies.
Parasitology	Scientific study of parasitism.*
Proteomics	The qualitative and quantitative comparison of proteomes under different conditions to further unravel biological processes.
Phycology	Branch of botany that deals with algae.*
Phylogeny	The evolutionary development and history of a species or higher taxonomic grouping of organisms.*
Physiology	Biological study of the functions of living organisms and their parts.*
Phytopathology	Science of plant diseases.
Population Ecology	Branch of ecology that studies the structure and dynamics of populations.
Population Genetics	Deals with evolution within species and is strongly dependent on mathematical models.
Quantitative Genetics	Study of continuous traits (such as height or weight) and its underlying mechanisms.
Sociobiology	Study of the biological determinants of social behavior, based on the theory that such behavior is often genetically transmitted and subject to evolutionary processes.*
Systems Biology	Field that seeks to integrate different levels of information to understand how biological systems function.
Taxonomy	The classification of organisms in an ordered system that indicates natural relationships.
Toxicology	Study of the nature, effects, and detection of poisons and the treatment of poisoning.
Virology	Study of viruses and viral diseases.
Xenobiology	A human doctor or biologist who is expert on the physiology of alien organisms and life forms.
Zoology	Branch of biology that deals with animals and animal life, including the study of the structure, physiology, development, and classification of animals.*

*American Heritage Dictionary of the English Language, 4th ed., 2000.

on what you put into it. Biology can intrigue and enlighten you or it will confound and frustrate you. It all depends upon the effort you are willing to put into your studies. To support that effort, this book helps sift through all the technical information and explains the value of understanding the basic concepts of biology that are relevant to everyday life and to our environment. If you keep an open mind, follow each lesson step by step, and complete the chapter review questions, this book will change the way you view the world and life within it.

As mentioned in the preface, *Biology for Nonbiologists* targets both a specific audience and a more general audience, including:

- Professionals (especially in environmental sciences) who need a quick review of biological principles and applications
- Nonbiologists who need an information source

- First-year biology students who need a study companion
- General readers who do not want to limit themselves to their own specific area of expertise but prefer to add knowledge of the basic tenets of biology to their repertoire

Key features of the text include:

- For the nonbiologist, familiarity with the jargon and key concepts of environmental biology as it relates to the environmental profession
- Interesting and up-to-date applications, with numerous examples and easy-to-follow, step-by-step solutions in the text
- Review of the scientific method
- Extensive lists of references and additional reading
- Easy-to-understand tables, figures, and diagrams
- Easy-to-understand language, with key points of caution/interest to help readers avoid misunderstanding or misapplication
- Common examples which allow the reader to understand the context of the information and its relevance to everyday life
- Explanations of concepts without mathematics and with little physical science
- For those entering the environmental profession, provides insights to suggest paths of inquiry in terms of career choices and goals

Why Study Biology?

In 1876, the English biologist and educator T.H. Huxley stated the following in his *Collected Essays*:

> It is my duty tonight to speak about the study of Biology, and while it may be that there are many of my audience who are quite familiar with that study, yet as a lecturer of some standing, it would, I know by experience, be very bad policy on my part to suppose such to be extensively the case. On the contrary, I must imagine that there are many of you who would like to know what Biology is; that there are others who have that amount of information, but would nevertheless gladly hear why it should be worth their while to study Biology; and yet others, again, to whom these two points are clear, but who desire to learn how they had best study it, and, finally, when they had best study it.

Huxley acknowledges that there are those who "would gladly hear why it should be worth their while to study Biology." Stated differently, why should we care about biology? Or, simply, why study biology?

In attempting to make my point, maybe I'm asking the wrong question. Instead, maybe the correct question is: "What other science discipline has proved more beneficial and exciting over the last century than biology? One need only review the great discoveries made in biology throughout the years that have granted amazing technological advances. These advances have enabled us, for example, to manufacture insulin cheaply using transformed bacteria, giving new hope to diabetic patients" (IUS 2006).

Moreover, many past fatal diseases are now successfully treated because of advances in biological science. "In addition to solving human problems, biology helps us to understand our environment and learn how to care for it" (IUS 2006). Biology has allowed us to develop new crop varieties. It has enabled us to genetically enhance crops to resist drought and pests ("reducing the need for pesticides and increasing productivity" [IUS 2006]), and increase their cold tolerance. Emerging computer technologies allow us to bridge the gap between biology and technology—creating new career opportunities. For all these reasons, and for others, we study biology.

"Aside from our constant desire for technological advances, biology appeals to that innate desire to satisfy human curiosity" (IUS 2006). Hiking a trail in a national park, for example, can be full of biological wonder and tragedy. "We wonder at the rapid wing-beat of a hummingbird or watch with interest as an army of ants drags seeds underground" (IUS 2006). The flora on the trail captivates us and brings us fully into its own world, shutting out all the other worlds of our lives. For a brief span of time along the trail, the office, cities, traffic, and the buzz and grind of work melt away into forest. These are the positive aspects of getting close to nature. A trained biologist would notice all of these things that please us. At the same time, however, a trained biologist would also notice those things that might not please us. Along the trail, for example, we notice standing dead Fraser fir and red spruce; stands of pollution-killed trees where fallen gray tree trunks crisscross each other in a horrible game of giant jackstraws; standing dead red spruce silhouetted by polluted fog; understories of brambles looking up at dead sugar maples; foliage areas bleached by ozone; trees of all varieties starved to death—the needed soil nutrients leached away by decades of acid deposition, and the trees weakened until they are no longer capable of withstanding the assaults of ordinary disease and bad weather; tropospheric ozone damage with chlorophyll bleaching; branch dieback on northern red oak; premature leaf-drop on quaking aspens; thinning crowns on sugar maples; tipped-over tulip poplars with rotted roots; chemically green ponds in areas where active strip mining occurs; an orange waterfall next to an abandoned mine; and a high overlook, where 25 years earlier the surrounding landscape was visible for 50 miles, now veiled in thick, stagnant, polluted fog with visibility reduced to 2 or 3 miles. A trained biologist would revel in nature's wonders and be horrified by the heavy hand of man. This is why we study biology (IUS 2006).

"This same innate feeling of curiosity would be experienced by the biological researcher who watches bacteria glow with the green florescent protein spliced into their DNA, who discovers the fossil of an ancient plant, or successfully sequences a gene responsible for nerve development. Biology is fascinating, beautiful, and complex, and biology is truly the science of life" (IUS 2006). This is why we study biology.

Biology can also be controversial. Current controversial aspects of biological research and findings include: family planning (birth control), how much money to spend on endangered species protection, biomedical research dealing with human fetal tissue, irradiating food to make it safe to eat, dangers involved in cloning animals and humans, and the perception of many that biologists are playing God. Some of these controversies remain within biology, others involve economic, moral, ethical, and religious considerations. Again, this is why we study biology.

The biologists learn from nature and the environment that surrounds and supports us. It is human nature to make nature serve us. Without biology (and the other sciences), we are at nature's mercy. To control nature (to the extent possible), we must learn its laws, and then use them responsibly.

Environmental professionals must learn the "laws" of biology and its uses, but they must know more. Environmental professionals must know the ramifications of biology when it is out of control. Biology properly used can perform miracles. Out of control, biologicals and their effects can be, and have been, devastating. Many of the Occupational Safety and Health Administration (OSHA) regulations and guidance documents dealing with biological safety and emergency response procedures resulted because of catastrophic events involving biologicals. For example, OSHA's Technical Manual, Section III: Chapter 7 originated because of the horrific results of the Legionnaires' disease incident in 1976. In the Legionnaires' incident, an outbreak of pneumonia caused 34 deaths at a 1976 American Legion convention in Philadelphia. The causal factor in the Legionnaires' incident, *Legionella pneumophila*, was first identified in 1977 by the Centers for Disease Control and Prevention (CDC). *L. pneumophila* had undoubtedly caused previous pneumonia outbreaks, but the organism's slow growth and special growth requirements prevented earlier discovery.

✓   **Important Point**: In the U.S., Legionnaires' disease is considered to be fairly common and serious, and the *Legionella* organism is one of the top 3 causes of sporadic, community-acquired pneumonia. Because it is difficult to distinguish this disease from other forms of pneumonia, many cases go unreported. Approximately 1,000 cases are reported annually to the CDC, but it is estimated that over 25,000 cases of the illness occur each year resulting in more than 4,000 deaths.

The ultimate causal bacterium of the 1976 Legionnaires' incident in Philadelphia was located in the air-conditioning ducts within the hotel and eradicated by several personnel trained in the biological sciences. Because of this incident, OSHA and CDC, with extensive guidance by biologists and other trained scientists, developed and implemented guidelines to prevent future incidents.

What Is Life?

We defined biology as the study of life and living organisms. Seems simple enough, but is it really that simple? Have you asked what life is? What does it mean to be alive? Have you ever tried to define life? If so, how did you define it? "If these questions strike you as odd, consider them for a moment (they are almost as hard as defining the origin of life). Of course we all have an intuitive sense of what life is," (UCAR 2000–2004) but if you had difficulty, as is probably the case, with answering these questions, you are not alone. These questions are open to debate and have been from the beginning of time. One thing is certain: life is not a simple concept and it is impossible to define.

Along with the impossibility of defining life definitively, "it is not always an easy thing to tell the difference between living, dead, and nonliving things. Prior to the 17th century, many people believed that nonliving things could spontaneously turn into living things. For example, it was believed that piles of straw could turn into mice. Obviously, that is not the case. There are some very general rules to follow when trying to decide if something is living, dead, or nonliving" (Utah State Office of Education 2000). "Scientists have identified 7 basic characteristics of life. Keep in mind that for something to be described as living, that something must display all 7 of these characteristics (i.e., "characteristic" is plural). Although many of us have many different opinions about what "living" means, the following characteristics were designated "characteristics of living things" with the consensus of the scientific community" (ThinkQuest Team C003763 2000).

- "**Living things are composed of cells**" (ThinkQuest Team C003763 2000). "Living things exhibit a high level of organization, with multicellular organisms being subdivided into cells, and cells into organelles, and organelles into molecules, etc." (Johnson 2006).
- "**Living things reproduce.** All living organisms reproduce, either by sexual or asexual means" (ThinkQuest Team C003763 2000).
- **Living things respond to stimuli.** "All living things respond to stimuli in their environment" (ThinkQuest Team C003763 2000).
- "**Living things maintain homeostasis.** All living things maintain a state of internal balance" (ThinkQuest Team C003763 2000) in terms of temperature, pH, water concentrations, etc.
- "**Living things require energy**" (ThinkQuest Team C003763 2000). "Some view life as a struggle to acquire energy (from sunlight, inorganic chemicals, or another organism) and release it in the process of forming adenosine triphosphate (ATP)" (Farabee 1992–2007). "The conventional view is that living organisms require energy, usually in the form ATP. They use this energy to carry out energy-requiring activities such as metabolism and locomotion" (ThinkQuest Team C003763 2000).
- "**Living things display heredity.** Living organisms inherit traits from the parent organisms that created them" (ThinkQuest Team C003763 2000).
- "**Living things evolve and adapt**" (ThinkQuest Team C003763 2000). All organisms have the ability to adapt or adjust to their surroundings. An example of this might be adapting to environmental change resulting in an increased ability to reproduce.

✓   **Interesting Point**: "Again, if something follows one or just a few of the characteristics listed above, it does not necessarily mean that it is living. To be considered alive, an object must exhibit *all* of the characteristics of living things" (Utah State Office of Education 2000).

"A good example of a nonliving object that displays at least 1 characteristic for living would be sugar crystals growing on the bottom of a syrup dispenser" (Utah State Office of Education 2000). On the other hand, there is a stark exception to the characteristics above. For example, mules cannot reproduce because they are sterile. Another nonliving object that exhibits many of the characteristics of life is a flame. Think about it, a flame:

- Is organized
- Reproduces
- Is irritable

- Respires
- Requires nutrition
- Excretes
- Grows
- Moves

We all know that a flame is not alive, but how do we prove that to the skeptic? The best argument we can make is:

1. Nonliving materials never replicate using DNA and RNA (hereditable materials).
2. Nonliving material cannot carry out anabolic metabolism.

DEFINITION OF KEY TERMS

After reviewing the characteristics of life listed above, it should be obvious that we need to follow Voltaire's advice, that is: "If you wish to converse with me, please define your terms." Therefore, in the following, many of the standard public domain terms (and cited others) used to describe the characteristics of life, or related to the characteristics of life, are defined (Quia 2007):

- **Anabolism**—the utilization of energy and materials to build and maintain complex structures from simple components.
- **Amino acids**—building blocks of proteins.
- **Asexual reproduction**—requires 1 parent cell.
- **Autotrophic**—using light energy (photosynthesis) or chemical energy (chemosynthesis).
- **Catabolism**—the breaking down of complex materials into simpler ones using enzymes and releasing energy.
- **Carbohydrates**—main source of energy for living things (e.g., sugar and starch).
- **Cellular basis of life**—cells are the basis of life. There are 2 types of cells: prokaryotes and eukaryotes.
- **Cold-blooded animals**—body temperatures change with the environment.
- **Compound**—2 or more elements chemically combined.
- **Digestion**—process by which food is broken down into simpler substances.
- **DNA**—the double helix of DNA is the unifying chemical of life; its linear sequence defines the diversity of living things.
- **Element**—pure substance that cannot be broken down into simpler substances
- **Enzymes**—proteins that regulate chemical activities.
- **Excretion**—process of getting rid of waste materials.
- **Eukaryotes**—these are cells with a nucleus; they are found in human and other multicellular organisms (plants and animals), also algae, protozoa.
- **Evolution**—the modification of species, is the core theme of biology.
- **Food**—needed by living things to grow, develop, and repair body parts.

- **Heterotrophic**—obtaining materials and energy by breaking down other biological material using digestive enzymes and then assimilating the usable by-products.
- **Ingestion**—taking in food.
- **Inorganic compounds**—generally, do not contain carbon.
- **Life span**—maximum length of time an organism can be expected to live.
- **Lipids**—energy-rich compounds made of carbon, oxygen and hydrogen.
- **Metabolism**—chemical reactions that occur in living things.
- **Movement**—nonliving material moves only as a result of external forces while living material moves as a result of internal processes at cellular level or at organism level (locomotion in animals and growth in plants).
- **Nucleic acids**—store information that helps the body make proteins it needs.
- **Organic compounds**—found in living things and contain carbon.
- **Organism**—any living thing.
- **Prokaryotes**—are cells without a nucleus. These include bacteria and cyanophytes (blue-green algae). The genetic material is a single circular DNA and is contained in the cytoplasm, since there is no nucleus.
- **Proteins**—used to build and repair cells; made of amino acids.
- **Respiration**—taking in oxygen and using it to produce energy.
- **Response**—action, movement, or change in behavior caused by a stimulus.
- **Sexual reproduction**—requires 2 parent cells.
- **Stimulus**—signal that causes an organism to react.
- **Structure and function**—at all levels of organization, biological structures are shaped by natural selection to maximize their ability to perform their functions.
- **Unity in diversity**—explained by evolution: all organisms are linked to their ancestors through evolution; scientists classify life on Earth into groups related by ancestry; related organisms are descended from a common ancestor and have certain similar characteristics.
- **Warm-blooded animals**—maintain a constant body temperature.

Levels of Organization

All living things are organized into several basic levels of organization. The levels of organization from the smallest to the largest are:

- **Atoms**—basic units of matter having a specific chemical property
- **Molecules**—smallest multi-atom-containing unit having a specific chemical property
- **Organelles**—include many subcellular structures, usually found only in eukaryotes
- **Cell**—basic unit of life
- **Tissue**—group of similar cells acting as a structural or functional unit
- **Organ**—group of dissimilar tissues that act together and have a specific function
- **Organ system**—group of dissimilar organs that together have a specific function
- **Organism**—contains cells and is capable of growing and replicating itself

In addition to the levels of organization for living things, there are also levels of organization in the biosphere. From the smallest to largest group these are:

- **Species**—a group of organisms that resemble one another closely
- **Population**—all organisms of the same species in a specific space and time
- **Community**—all populations in a specific space and time
- **Ecosystem**—community plus abiotic environment in a specific space and time
- **Biome**—a major regional or global biotic community, such as a grassland or desert, characterized chiefly by the dominant forms of plant life and the prevailing climate
- **Biosphere**—part of a planet's outer shell—including air, land, surface rocks, and water—within which life occurs, and which biotic processes in turn alter or transform

Major Theories of Biology

"Modern biology is based on several great ideas, or theories" (Farabee 1992–2007):

- Evolution by natural selection theory
- Inheritance theory
- Cell theory
- Biological classification
- Bioenergetics (the energy that powers life)
- Homeostasis
- Ecosystems

Evolution by natural selection theory has 2 parts: (1) Species change over generations, and (2) natural selection is the mechanism for change. Many scientists accept this as biology's most important theory. "However, it remains a lightning rod for politicians, television preachers, and school boards. Much of this confusion results from what the theory says and what it does not say" (Farabee 1992–2007). Basically, we don't know what we don't know about evolution.

Inheritance theory is intimately connected to evolution and explains how traits are inherited by offspring from parents. Mendel's laws or principles of inheritance apply to all organisms. Traits are passed from one generation to the next via genes (hereditary factors) (DiResta 2001).

Cell theory was proposed in 1838 by Schleiden and Schwann; their proposal changed cell biology research forever. The cell theory states that:

1. All life-forms are made from 1 or more cells.
2. Cells only arise from pre-existing cells.
3. The cell is the smallest form of life.

Biological classification classifies living organisms by similarities and differences (DiResta 2001).

Bioenergetics refers to the flow and transformation of energy in and between living organisms and between living organisms and their environment.

Homeostasis is the maintenance of a dynamic range of internal conditions (maintained in narrow limits) within which the organism can function.

Ecosystems make up communities and their environments. The ecosystem concept recognizes that organisms do not exist alone.

The Scientific Method

Note: Much of the information in the following section is from USEPA's *What Is the Scientific Method* (www.epa.gov/maia/html/scientific.html).

In order to conduct science, one must know the rules of the game. The word "science" is derived from a Latin verb meaning "to know." Scientists use the *scientific method* to construct an accurate representation of the world through the testing of scientific theories. These theories enhance our understanding and knowledge of the world. A theory, such as Einstein's theory of relativity or Darwin's theory of natural selection, accounts for many facts and attempts to explain a great variety of phenomena. Such a unifying theory does not become widely accepted in science unless its prediction can withstand thorough and continuous testing by experiments and observations (USEPA 2006; Campbell 2004).

It is important to point out that the scientific method is to be used as a guide that can be modified. In some sciences, such as geology, and taxonomy, lab experiments are not necessarily performed. Instead, after formulating a hypothesis, additional observations and/or collections are made from different localities.

Sir Karl Raimund Popper (1902–1994) developed the theory of the scientific method in 1934 in his book *The Logic of Scientific Discovery*. The U.S. Environmental Protection Agency (2006) points out that many scientists including biologists, chemists, ecologists, physicists and environmental practitioners all use the scientific method to test new theories.

While there are several forms of the scientific method, USEPA (2006) points out that the basic steps involved are:

1. Initial observations and objectives
2. Hypothesis formulation
3. Data collection (e.g., record observations)
4. Analysis of the data (e.g., perform calculations) to test the hypothesis
5. Summarization of results (create tables, graphics, etc.)
6. Discussion of limitations and conclusions
7. Identification of future research needs

The University of Maryland School of Medicine (Department of Pathology) suggests an ideal example of the scientific method and its importance. A researcher may observe that a large number of insects all have 3 pairs of legs including flies, beetles, grasshoppers and wasps. A conclusion may be drawn that all insects have 3 pairs of legs. Then after evaluating additional insects including cockroaches, crickets, moths and bees, a hypothesis might be formed that all insects have 3 pairs of legs. However, a good scientist would not stop there. New hypotheses should be formed to further test the initial observation. Perhaps immature moths should be considered, as they are insects too, so they should have 3 pairs of legs. However, findings would determine that caterpillars or immature moths do not have any legs. Then the generalization becomes reformulated into all adult insects have 3 pairs of legs.

HOW IS THE SCIENTIFIC METHOD USED?

1. **Initial observations and objectives**—Scientists are usually curious about their surroundings and may notice something and want to understand more about it. This type of curiosity may lead them to observe their surroundings more carefully. As a result, when they document their observations, questions may arise which are then formulated into hypotheses.
2. **Hypothesis formulation**—As observations are made, questions are formulated and scientists try to answer these questions, which leads to guesses or hypotheses. If observations do not support these statements, the hypothesis is rejected. A hypothesis must be stated in a way that can be tested by the scientific method. Reviewing similar studies performed by others can also be helpful in this step.
3. **Data collection**—In order to test the hypotheses, data must be collected. Scientists must design a data collection plan which considers what to collect, when to collect, and how many samples are necessary. Developing sample surveys involves determining appropriate sample size, monitoring frequency and the need for repetition. Through the use of variables and controls, results can be determined and documented. Variables are those factors being tested in an experiment, which are usually compared to a control. A control is a known measure to which scientists can compare their results.

   The scientist must also determine which equipment, supplies or materials are necessary to complete the study. Prior to conducting an experiment it is important to document data collection methods. This step will ensure the quality of the experiment if someone else should reproduce it. Most scientific papers published in journals include a methods section documenting the way in which the experiment was performed. Attention must be paid to make sure that data collection methods are kept unbiased. Data can be represented by many formats. For example:

   > The Environmental Monitoring and Assessment Program Near Coastal (EMAP-NC) Program Plan for 1990: Estuaries (EPA/600/4-90/033) provides an example of applying the scientific method. Initially, observations were made regarding estuarine and coastal ecological condition leading to the identification of the problem—humans. From this, a solution to the problem was proposed—the need for better environmental surveillance to assess the status and trends of the nation's ecological problems. To achieve this, a sampling design was developed. Since the problem identified estuarine and coastal wetlands, estuaries, coastal waters, and the Great Lakes as areas of potential concern, the scope of the

project needed to be determined. Due to budget and time limitations, the EMAP program chose to focus their initial efforts on the state of estuaries along the nation's coastline (e.g., where to collect). The sampling design for the EMAP-NC had 3 major elements including a regionalization scheme, a classification scheme, and a statistical design.

Specifically, the statistical design identified what parameters were to be sampled, when sampling should occur, and how many samples should be taken. Parameters to be sampled were identified including, but not limited to benthic and fish species composition and biomass, habitat indicators including salinity, temperature, pH, dissolved oxygen, sediment characteristics, and water depth. Station locations were chosen using a randomly placed systematic grid and sampling times were identified to fall within an index period, which was chosen to be the summer. (EPA 2006)

4. **Analysis of the data**—This step is necessary to prove or disprove a hypothesis by experimentation. The methods involved in testing/analyzing the data are also important since an experiment should be repeated by others to ensure the quality of results. For instance, if 2 people on different sides of the country decide to perform the same experiment, they should end up with the same results. Statistics are then used to analyze the data. Descriptive statistics are a means of summarizing observational data through the calculation of a mean, mode, average, standard deviation, variance, etc. More advanced comparisons can also be completed. This step is very important as it transforms raw data into information, which can be used to report results in a user-friendly format. To further analyze the data, parametric or nonparametric statistical tests can be performed. Parametric tests assume that data are normally distributed. Generally, to ensure that a parametric test is appropriate, tests for normality and variance need to be completed to identify any necessary transformations (e.g., taking the log of the values) or if a nonparametric test is warranted. Nonparametric tests make fewer assumptions about the distributions of the data.

5. **Summarization of results**—The presentation of the results is very important. Often scientists will rely heavily on graphics, tables, flow charts, maps, and diagrams to facilitate the interpretation of the results. Graphics can be used to model future predictions. Graphics (e.g., scatter plots) can also assist with the identification of relationships (e.g., correlations) between environmental parameters. If 2 variables are correlated, when one changes the other will do so in a related manner. This relationship can be either positive or negative. Often when a correlation is found, it is assumed that there is a "cause and effect" relationship between the variables. Although there should be some logical basis for relating variables, cause is not demonstrated with a statistical technique. When we say 2 variables are correlated, we can say that they are associated in some way. A written discussion documenting identifiable trends or correlations generally accompanies these graphics. This step is just the presentation of results; it does not include any interpretation.

6. **Discussion of limitation and conclusions**—This is the section where the hypothesis is accepted or rejected. Many scientists no longer try to define cause and

effect parameters, but instead identify relationships between the data. In this manner, ideas can be formed about why certain results were found while identifying previous studies that may have had similar or contradicting results. It is important to reference all studies so that other scientists can refer to them if necessary.

7. **Identification of future research needs**—This may include areas of related interest that should be studied to better understand the subject. This section may give information about limitations of the study, such as what items should be modified to try to reach the intended goal.

✓   **Important Point**: In addition to the 7 steps involved, there are 3 important factors to consider. First, throughout the entire process of the scientific method, quality assurance/quality control (QA/QC) is extremely important. The goal of QA/QC is to ensure that environmental data are of sufficient quantity and quality to support the intended use of the data. Second, methods and observations, statistical procedures, and results should be documented and reviewed to ensure a quality study. This step is vital to the third factor—the repeatability of the process. If 2 exact experiments are conducted, they should have the same findings. Additional testing ensures the quality of the results. Even if the experiment does not prove the hypothesis, it is still important. Having the knowledge that an experiment did not work is a step toward finding the answer.

WHY IS THE SCIENTIFIC METHOD IMPORTANT?

The scientific method has played an instrumental part in scientific research for almost 500 years. From Galileo's experiment back in the 1590s to current scientific research, the scientific method has contributed to the creation of vaccines and advancements in medicine and technology. Through the use of the scientific method, scientific theories can be tested. A scientific theory is a logical explanation of observed events. Once a scientific method has been tested and widely accepted as true, it becomes scientific law (e.g., Newton's law of gravity). However, scientific methods must be continually examined for possible errors. Openness to new innovative ideas and an organized approach of skepticism are necessary to protect against collective bias into scientific results. The manner in which the hypothesis is formed, as well as the way data is collected, analyzed, and interpreted all need to be monitored for the potential introduction of bias.

The scientific method also ensures the quality of data for public use. This can be completed through submitting work for a peer review. A peer review is a critical review by technical experts without a vested interest in the particular investigation. Peer reviews confirm that the research has been conducted in a scientifically sound manner. For example, peer reviews are performed on all scientific articles being submitted to journals to check the quality of each study before it is released to the public. The EPA always peer reviews its studies and data before they are released into the public domain.

Case 1.1. Scientific Method: Pfiesteria and Fish Health

Note: The following is adapted from a Sea Grant Maryland (SGM)—funding provided by the National Oceanic and Atmospheric administration (NOAA), work titled: *The Scientific Method, Fish Health and Pfiesteria* (related to fish health in the Chesapeake Bay) with the writings of William T. Keeton (1996) and a presentation by David Goshorn, Ph.D. (2006).

Goshorn (2006) points out that:

> *Pfiesteria* is a very small, single-celled organism without flagella. It's got an extremely complex life cycle. Most of the time populations are benign, feeding on algae and bacteria, but some populations, not all, are capable of producing a toxin which can cause fish health problems and, apparently, human health problems. In 1997, Maryland experienced four separate toxic outbreaks on three different Eastern Shore rivers. North Carolina has had its problems for quite some time.</ext>

Sea Grant Maryland (2006), in their *Fish Disease Information: The Scientific Method, Fish Health and Pfiesteria* reported (in regards to fish health in the Chesapeake Bay) that their "understanding of the situation is that fish have had lesions and that fish have died. *Pfiesteria piscicida* and *Pfiesteria*-like organisms have been cultured from water samples taken in the vicinity where fish with lesions have been observed or where fish kills have occurred. It was reported that *Pfiesteria*-like organisms have caused lesions and mortalities in laboratory exposures."

Well, sounds like the jury is out; a simple matter of cause and effect. *Pfiesteria* were present, fish had lesions, and fish died. Moreover, *Pfiesteria*-like organisms have caused lesions and mortalities in lab exposures. That seals the deal; a slam duck verdict. Right? Not so fast my environmental practitioner friends. We know better (hopefully) than to jump the gun, don't we?

As SGM points out "as good scientists know, it's not that easy."

Notwithstanding that many of us are scientists or want-to-be scientists or those who just want to know more about science, there are some who might ask: "Why, in this case, is it not easy to conclude that fish lesions are caused by *Pfiesteria* or *Pfiesteria*-like organisms?" SGM points out that the "very same water which was cultured for *Pfiesteria* could have also grown bacteria." Therefore, in order to properly study the cause of the fish lesions and the possibility that *Pfiesteria* are culprits, we need to follow Koch's postulates.

Robert Koch (1882; 1884; 1893), a German physician and bacteriologist, in the course of his studies of anthrax and tuberculosis, formulated 4 postulates (i.e., 4 criterions for judging whether a given bacteria is the cause of a given disease) in 1884 and refined and published them in 1890. To say that Koch's criteria brought some much-needed scientific clarity to what was then a very confused field is an understatement. Koch's postulates are still used today.

Koch's postulates are as follows:

1. The organism must be found in all animals suffering from the disease, but not in healthy animals.
2. The organism must be isolated from a diseased animal and grown in pure culture.
3. The cultured organism should cause disease when introduced into a healthy animal.
4. The organism must be reisolated from the experimentally infected animal.

It should now be clear why it is difficult to establish a cause and effect relationship between fish lesions and *Pfiesteria*. SGM (2006) points out "*Pfiesteria* (or its toxin) has not yet

Case 1.1.   Scientific Method:
Pfiesteria and Fish Health (*continued*)

been isolated from fish or fish lesions." The record shows that fish with lesions have been taken from waters where *Pfiesteria*-like organisms have not been identified (of course that does not mean that they were not present). Moreover, based on tests, it has been shown that lesions on fish from lab exposures are not identical to those seen in some natural habitat-collected specimens. This, especially the non-specificity of the lesions, means that the lesions could have been caused by a number of different causative agents (e.g., bacteria, virus, etc.). SGM reports that they were unable to isolate the infectious organism from the host, culture it, and reinfect another host to be able to observe if the same lesions occur.

So, does the above account, using Koch's postulates, prove or disprove the association of *Pfiesteria* and fish lesions? At the present time we cannot definitively say for sure one way or other on the *Pfiesteria*-fish lesion connection. Instead, we are at the hypothesis stage and testing is ongoing, a work in progress. The point is that eventually, after the hypothesis is tested, challenged, revised, retested, rechallenged, and re-revised, a theory is developed. This theory still remains subject to debate by all who question it; they may counter it with new evidence. Ultimately, the theory may be accepted as fact.

 Important Point: It is important to point out that Koch's postulates have their limitations and so may not always be the last word (Med Net, 2006). They may not hold if:

1. The particular bacteria (such as the one that causes leprosy) cannot be "grown in pre-culture" in the lab.
2. There is no animal model of infection with that particular bacterium.

Text Themes

Every topic presented in this book will relate to one or more of the following themes:

- All living things are made of cells, which carry out the basic functions of living organisms.
- The way living organisms are made facilitates the way they work.
- In living systems, matter cycles and energy flows.
- Living organisms have the ability to regulate what is going on internally despite what is going on externally.
- Reproduction is necessary for the survival of any species and is accomplished by the passing on of hereditary information.
- Within the diversity of living organisms there are fundamental similarities.
- Scientists assume that the universe is a system that can be understood by careful systematic study.

Chapter Review Questions

1. What is biology?
2. List the 7 characteristics of life.
3. What are the building blocks of proteins?
4. Special types of proteins that regulate chemical activities are: _____.
5. _____ store information that helps the body make the protein it needs.
6. List the levels of organization of living things from smallest to largest.
7. List the levels of organization in the biosphere from smallest to largest groups.
8. Define cell theory.
9. List the basic steps of the scientific method.
10. List Koch's postulates.

THOUGHT-PROVOKING QUESTION

1. Are viruses alive?

References and Additional Reading

American Heritage Dictionary of the English Language, 4th ed., 2000. Houghton Mifflin Company.

Bordenstein, S. 2006. *Microbial Life: Educational Resources*. Accessible by http://serc.carleton.edu/microbelife/k12/bioinformatics/index.html.

Campbell, N.A. 2004. *Biology: Concepts & Connections*, 4th CD-ROM ed. Benjamin-Cummings Publishing Company.

DiResta, D. 2001. *Why Study Biology?* Accessible by http://fig.cox.miami.edu/~ddiresta/bil101/Lec01.htm.

Farabee, M.J. 2007. *An On-Line Biology Book*. Accessible by www.emc.maricopa.edu/faculty/farabee/biobk/BioBookToc.html.

Goshorn, D. 2006. Proceedings—DELMARVA Coastal Bays Conference III: Tri-State Approaches to Preserving Aquatic Resources. USEPA.

Huxley, T.H. 1876. *Science & Education, Volume III, Collected Essays*. D. Appleton & Company.

IUS. 2006. *Biology—The Science of Life*. Indiana University Southeast. Accessed June 9, 2006, at www.ius.edu/biology/whybio.stm.

Johnson, J.G. 2006. *The Science of Life*. Accessible by www.sirinet.net/~jgjohnso/studyoflife.html.

Jones, A.M. 1997. *Environmental Biology*. New York: Routledge.

Keeton, W.T. 1996. *Biological Science*. R.S. Means Company.

King, R.M. 2003. *Biology Made Simple*. New York: Broadway Books.

Koch, R. 1882. Über die Ätiologie der Tuberkulose. In *Verhandlungen des Kongresses für Innere Medizin*, Erster Kongress, Wiesbaden.

Koch, R. 1884. *Mitt Kaiser Gesundh*. 2, 1–88.

Koch R. 1893. *J. Hyg. Inf.* 14, 319–333.

Larsson, K.A. 1993. Prediction of the pollen season with a cumulated activity method. *Grana*, 32, 111–114.

MedTerms Medical Dictionary. 2006. s.v. "Koch's Postulates," MedicineNet.com. Accessed June 9, 2006, at www.medterms.com/script/main/hp.asp.

Quia. 2007. *Characteristics of Life.* Accessed August 8, 2006, at www.quia.com/jg/924list.html.

SGM. 2006. *The Scientific Method, Fish Health and Pfiesteria.* University of Maryland; NOAA.

Spellman, F.R., and N.E. Whiting. 2006. *Environmental Science and Technology*, 2nd ed. Rockville, MD: Government Institutes.

Spieksma, F.T. 1991. Aerobiology in the nineties: aerobiology and pollinosis. *International Aerobiology Newsletter* 34, 1–5.

ThinkQuest Team C003763. 2000. *Characteristics of Life.* Accessible by http://library .thinkquest.org/C003763/index.php?page=origin06.

USEPA. 2006. *What Is the Scientific Method?* Accessed June 8, 2006, at www.epa.gov/maia/ html/scientific.html.

Utah State Office of Education. 2000. *Characteristics of Living Things.* Accessible by www.usoe .k12.ut.us/CURR/science/sciber00/7th/classify/living/2.html.

Windows to the Universe team. *What Is Life.* Boulder, CO: 2000–2004 University Corporation of Atmospheric Research (UCAR) 1995–1999, 2000 The Regents of the University of Michigan, April 29, 2004. Online. Available: www.windows.ucar.edu.

CHAPTER 2

Chemistry

Verily, chemistry is not a splitting of hairs when you have got half a dozen raw Irishmen in the laboratory.

—Henry David Thoreau (1817–1862)

Topics in This Chapter

- What Is Chemistry?
- What Is Matter?
- Elements and Compounds
- Atomic Structure
- Chemical Bonding
- Water Chemistry
- Acids, Bases, pH
- Organic Chemistry
- Macromolecules

Why do biologists need to study chemistry? In a general sense, consider that, on foundational levels, chemistry affects everything we do. Not a single moment of time goes by during which we are not affected in some way by a chemical substance, chemical process, or chemical reaction. Chemistry affects every aspect of our daily lives; this is why we study chemistry (Spellman and Whiting 2006).

Moreover, it is important to point out that biology is a very large field. When we think of educating biologists, we think about people who work in everything ranging from the evolution of birds' beaks to the chemical structure of mitochondria. The fact is that much of biology has become chemical in nature. Biology students and practitioners of the biological sciences need a strong background in chemistry. This is why we study chemistry.

In a specific sense, consider that almost every environmental pollution problem we face today (and probably tomorrow) has a chemical basis. In short, biological studies conducted by environmental practitioners, to examine critical environmental problems—including stream pollution, groundwater contamination, and insect infestation of our forests—would be difficult, if not impossible, without some fundamental understanding of chemistry. And of course, a biologist whose primary responsibility is to protect the environment must be well-grounded in chemical principles and the techniques of chemistry in general, because many of these techniques are used by biologists to solve environmental problems. Again, this is why we study chemistry.

Though unlikely for the modern practicing biologist, the general reader who uses this text may not necessarily have some fundamental knowledge of chemistry. In regards to the practicing biologist, this text provides a review of chemistry basics. This chapter's topics were selected with the goal of reviewing or providing only the essential chemical principles required to understand the nature of biology and the environmental problems we face, and the chemistry involved in scientific and technological approaches to their solutions.

✓   **Important Point:** Ninety-two elements are found naturally in the Earth's crust. Of these, only 11 are common in living organisms but many others are important despite occurring only in trace quantities.

What Is Chemistry?

Chemistry is the science concerned with the composition of matter (gas, liquid, or solid) and of the changes that take place in it under certain conditions. The following is a comprehensive definition of chemistry provided by *The American Heritage Dictionary*.

> chem•is•try n. **1**. the science of the composition, structure properties, and reactions of matter, especially of atomic and molecular systems. **2**. the composition, structure, properties, and reactions of a substance.

Every substance, material, and/or object in the environment is either a chemical substance or mixture of chemical substances. As mentioned, your body is made up mainly of 11 elements and literally thousands of chemical mixtures. Moreover, the food we eat, the clothes we wear, the fuel we burn, and vitamins we take in form of natural or synthetic sources are all products of chemistry, wrought either by the forces of nature or the hand of man. Living things are made up of a profusion of organic chemicals. Chemical reactions occur all around us. For example, chemical reactions occur in health care, cooking, cosmetics, and automobiles. Complex chemical reactions involving carbon-based molecules take place constantly in every cell in our bodies, for example, from digestion to reproduction. Chemistry is about matter; its actual makeup, constituents, and consistency. It is about measuring and quantifying matter. To understand these processes we must journey to the world of matter, elements, compounds, physical and chemical changes, and atoms.

What Is Matter?

Matter is anything that takes up space and has mass. "All matter can exist in three states: gas, liquid, or solid. It is composed of minute particles termed *molecules*, which are constantly moving, and may be further divided into *atoms*. Molecules that contain atoms of one kind only are known as **elements**: those that contain atoms of different kinds are called **compounds**.

"Chemical compounds are produced by a chemical action that alters the arrangements of the atoms in the reacting molecules. Heat, light, vibration, catalytic action, radiation, or pressure, as well as moisture (for ionization), may be necessary to produce a chemical change. Examination and possible breakdown of compounds to determine their components is *analysis*, and the building up of compounds from their components is *synthesis*. When substances are brought together without changing their molecular structures, they are said to be *mixtures*" (Research Machine 2006).

✓   **Important Point**: Organic substances consist of virtually all compounds that contain carbon. All other substances are inorganic substances.

Elements and Compounds

A pure substance is a material from which all other materials have been separated. Pure substances (samples of pure substances are indistinguishable from each other, no matter what procedures are used to purify them or their origins) include copper metal, aluminum metal, distilled water, table sugar, and oxygen. All samples of pure table salt are alike and indistinguishable from all other table salt samples.

Usually expressed in terms of percentage by mass, a substance is characterized as a material having a fixed composition. Distilled water, for example, is a pure substance consisting of approximately 11% hydrogen and 89% oxygen by mass. By contrast, a lump of coal is not a pure substance because its carbon content may vary from 35% to 90% by mass. Materials (like coal) that are not pure substances are mixtures. When substances can be broken down into two or more simpler substances, they are called compounds.

When substances cannot be broken down or decomposed into simpler forms of matter, they are called elements. As mentioned, the elements are the basic substances of which all matter is composed. Salt is made up of the elements sodium and chloride. Water is made up of the elements hydrogen and oxygen.

You might ask, for example, is pure table salt a compound? *Compounds* are pure substances made of atoms of two or more elements chemically combined in fixed ratios. Table salt is not a compound because it is a mixture of sodium chloride, dextrose and calcium silicate (to keep it from clumping in high humidity) and potassium iodide (an important nutrient necessary for proper thyroid function). However, if these components are separated physically, each can be obtained in pure form and each meets all the criteria for compounds.

Table 2.1.   Elements Making Up 99% of Earth's Crust, Oceans, and Atmosphere

Element	Symbol	% Composition	Atomic Number
Oxygen	O	49.5%	8
Silicon	Si	25.7%	14
Aluminum	Al	7.5%	13
Iron	Fe	4.75%	20
Calcium	Ca	3.4%	20
Sodium	Na	2.6%	11
Potassium	K	2.4%	19
Magnesium	Mg	1.9%	12
Hydrogen	H	1.9%	1
Titanium	Ti	0.58%	22

At the present time there are only 100+ known elements, but there are well over a million known compounds. Of the 100+ elements, only 88 are present in detectable amounts on Earth, and many of these 88 are rare. Ten elements make up approximately 99% by mass of the Earth's crust, including the surface layer, the atmosphere, and the bodies of water (see Table 2.1). From the table, we see that the most abundant element on Earth is oxygen, which is found in the free state in the atmosphere, as well as in combined form with other elements in numerous minerals and ores.

Table 2.1 also lists the symbols and atomic number of the 10 chemicals listed. The symbols consist of either one or two letters, with the first letter capitalized. The atomic number of an element is the number of protons in the nucleus.

Only 11 of the elements found naturally in the Earth's crust are common to living organisms (see Table 2.2).

CLASSIFICATION OF ELEMENTS

Each element may be classified as a metal, nonmetal, or metalloid. Metals—elements that are typically lustrous solids—make up approximately 75% of the elements on the

Table 2.2.   The Main Life Elements and Their Main Roles in Living Organisms

Element	Chemical Symbol	Main Role(s)
Oxygen	O	cellular respiration, component of water
Calcium	Ca	skeletal agent/cell adhesion/muscle contraction
Sodium	Na	main positive ion bathing cells/nerve function
Potassium	K	main positive ion in cells/nerve function
Magnesium	Mg	component of many enzymes
Hydrogen	H	electron carrier/component of water and most organic molecules
Phosphorus	P	nucleic acids/important in energy transfer (ATP system)
Carbon	C	basis of all organic molecules
Sulfur	S	component of most proteins
Chlorine	Cl	main negative ion bathing cells
Nitrogen	N	component of all proteins and nucleic acids

periodic table. They are good conductors of heat and electricity, melt and boil at high temperatures, possess relatively high densities, and are normally malleable (can be hammered into sheets) and ductile (can be drawn into a wire). Examples of metals are copper, iron, silver, and platinum. Almost all metals are solids at room temperature (none is gaseous), the one exception being mercury, which at room temperature exists in a liquid state.

✓   **Interesting Point**: Metals easily lose electrons and corrode easily. They usually show a reaction with acids.

The 17 elements that do not possess the general physical properties just mentioned are poor conductors of heat and electricity, boil at relatively low temperatures, do not possess a luster, and are less dense than metals. These are called *nonmetals*. At room temperature, most nonmetals are either solids or gases (the exception is bromine—a liquid). Nitrogen, oxygen, and fluorine are examples of gaseous nonmetals, while sulfur, carbon, and phosphorus are examples of solid nonmetals.

✓   **Interesting Point**: Nonmetals tend to gain electrons.

The 6 *metalloids* have properties resembling both metals and nonmetals. The metalloids are boron, silicon, germanium, arsenic, tellurium, antimony, and polonium.

PHYSICAL AND CHEMICAL CHANGES

Internal linkages among a substance's units (between one atom and another) maintain the substance's constant composition. These linkages are called *chemical bonds*. When a particular process occurs that involves the making and breaking of these bonds, we say that a *chemical change* or *chemical reaction* has occurred. Combustion and corrosion are common examples of chemical changes that impact our environment.

Let's briefly consider a couple of examples of chemical change. A chemical change means that the reacting substance(s) are changed into new substances. The actual atoms involved remain, they are simply rearranged. The rearrangement is called a chemical reaction. When a flame is brought into contact with a mixture of hydrogen and oxygen gases, a violent reaction takes place. The covalent bonds in the hydrogen (H_2) molecules and of the oxygen (O_2) moles are broken and new bonds are formed to produce molecules of water, H_2O. This illustrates the process whereby chemical bonds are broken and new ones are made.

When mercuric oxide, a red powder, is heated, small globules of mercury are formed and oxygen gas is released. This mercuric oxide is changed chemically to form molecules of mercury and molecules of water.

✓   **Key Point**: Whenever chemical bonds are broken, or formed, or both, a chemical change takes place. The hydrogen and oxygen undergo a chemical change to produce water, a substance with new properties. A process like grinding some salt crystals into a fine powder does not involve the breaking of chemical bonds and the formation of new ones, so it is a physical change.

By contrast, a *physical change* (nonmolecular change) is one in which the molecular structure of a substance is not altered. That is, a physical change is any change not involving a change in the substance's chemical identity. When a substance freezes, melts, or changes to vapor, the composition of each molecule does not change. For example, ice, steam, and liquid water all are made up of molecules containing 2 atoms of hydrogen and 1 atom of oxygen. A substance can be ripped, or sawed into small pieces (and sawdust), or ground into powder, or molded into a different shape without changing the molecules in any way.

✓   **Interesting Point**: Iron (and other metals) can be made to be magnetic. This change in no way affects the chemical identity of the element. Iron that is magnetized rusts just as easily as iron that is not magnetized.

The types of behavior that a substance exhibits when undergoing chemical changes are called its *chemical properties*. The characteristics that do not involve changes in the chemical identity of a substance are called its *physical properties*. All may be distinguished from one another by these properties, in much the same way as certain features (DNA, for example) distinguish one human being from another.

Atomic Structure

All matter, whether liquid, solid, or gas is made up of atoms, and all atoms have the same basic structure. For example, if a small piece of an element, say copper, is hypothetically divided and subdivided and subdivided into the smallest piece possible, the result would be one particle of copper. This smallest unit of the element that is still representative of the element is called an atom.

Although infinitesimally small, the atom is composed of particles, principally electrons, protons, and neutrons. The simplest atom possible consists of a *nucleus* having a single *proton* (positively charged particle) with a single *electron* (negatively charged particle) traveling around it—we say that an atom of hydrogen has an atomic weight of 1 because of its single proton. The *atomic weight* of an element is equal to the total number of protons and *neutrons* (neutral particles) in the nucleus of an atom of an element. Electrons and protons bear the same magnitude of charge, but have opposite polarity.

The hydrogen atom (an atom of the chemical element hydrogen) also has an atomic number of 1 because of its single proton. The atomic number of an element is equal to the number of protons in its nucleus. A neutral atom has the same number of protons and electrons, so the atomic number is also equal to the number of electrons in the atom. The number of neutrons in an atom is always equal to or greater than the number of protons except in the atom of hydrogen.

The protons and neutrons of an atom reside in the nucleus. Electrons reside primarily in designated regions of space surrounding the nucleus, called *atomic orbitals* or

Table 2.3.   Subatomic Particles

Name	Symbol	Charge
neutron	n	0
proton	p	+1
electrons	e	−1

electron shells. Only a prescribed number of electrons may reside in a given type of electron shell. With the exception of hydrogen (with only 1 electron), 2 electrons are always close to the nucleus, in an atom's innermost electron shell. In most atoms, other electrons are located in electron shells some distance from the nucleus (see Table 2.3).

While neutral atoms of the same element have an identical number of electrons and protons, they may differ by the number of neutrons in their nuclei. Atoms of the same element having different numbers of neutrons are called *isotopes* of that element.

PERIODIC CLASSIFICATION OF THE ELEMENTS

Through experience, scientists have discovered that the chemical properties of the elements repeat themselves. Chemists summarize all such observations in the *periodic law*: the properties of the elements vary periodically with their atomic numbers.

In 1869, Dimitri Mendeleev, using relative atomic masses, developed the original form of what today is known as the periodic table, a chart of elements arranged in order of increasing proton number to show the similarities of chemical elements with related electronic configurations. The elements fall into vertical columns known as *groups.* Going down a group, the atoms of the elements all have the same outer shell structure, but an increasing number of inner shells. Traditionally, the alkali metals are shown on the left of the table and the groups are numbered IA to VIIA, IB to VIIB, and 0 (for noble gases). Now we more commonly classify all the elements in the middle of the table as transition elements and regard the nontransition elements as main-group elements, numbered from I to VII, with the noble gases in group 0.

Horizontal rows in the table are *periods.* The first 3 are called *short periods*; the next 4 (which include transition elements) are *long periods.* Within a period, the atoms of all the elements have the same number of shells, but with a steadily increasing number of electrons in the outer shell.

✓ **Interesting Point**: There are 7 periods and 8 groups in Mendeleev's periodic table. There were no 0 groups in Mendeleev's periodic table. Mendeleev left some gaps in his periodic table for elements to be discovered later on. He even predicted the properties of these yet to be discovered elements by studying the properties of the neighboring elements.

The periodic table is an important tool for learning chemistry because it organizes, tabulates, and presents a variety of information in one spot. For example, we can immediately determine the atomic number of the elements because they are tabulated

Figure 2.1.   The element sodium as it is commonly shown in one of the horizontal boxes in the periodic table.

11	Atomic Number
Sodium	Name
Na	Symbol

in the periodic table (see Figure 2.1). We can also readily identify which elements are metals, nonmetals, and metalloids. Usually a bold zigzag line separates metals from nonmetals, while those elements lying to each immediate side of the line are metalloids. Metals fall to the left of the line, and nonmetals fall to the right of it.

MOLECULES AND IONS

When elements other than noble gases (which exist as single atoms) exist in either the gaseous or liquid state of matter at room conditions, they consist of units containing pairs of like atoms. These units are called molecules. For example, we generally encounter oxygen, hydrogen, chlorine, and nitrogen as gases. Each exists as a molecule having two atoms. These molecules are symbolized by the notations O_2, H_2, Cl_2, and N_2, respectively.

The smallest particle of many compounds is also the molecule. Molecules of compounds contain atoms of 2 or more elements. The water molecule, for example, consists of 2 atoms of hydrogen and 1 atom of oxygen (H_2O). The methane molecule consists of 1 carbon atom and 4 hydrogen atoms (CH_4).

Not all compounds occur naturally as molecules. Many of them occur as aggregates of oppositely charged atoms or groups of atoms called *ions*. Gaining or losing some of their electrons charges atoms. Atoms of metals that lose their electrons become positively charged and atoms of nonmetals that gain electrons become negatively charged.

Chemical Bonding

In the 19th century, the then-known elements were arranged according to chemical bonding, recognizing that one group (the inert gases or noble gases) tended to occur in elemental form (i.e., not in a molecule with other elements). It was later determined that this group had outer electron shells containing 2 (e.g., helium) or 8 (e.g., neon, xenon, radon, krypton, etc.) electrons.

Generally, for the atoms likely encountered in biological systems, atoms tend to gain or lose their outer shell electrons to achieve a noble gas outer electron shell configuration of 2 or 8 electrons. The number of electrons that are gained or lost is characteristic for each element, and ultimately determines the number and types of chemical bonds that atoms of that element can form.

When compounds form, the atoms of one element become attached to, or associated with, atoms of other elements by forces called *chemical bonds*. Chemical bonding is a strong force of attraction holding atoms together in a molecule. Various types of chemical bonds occur. Transfer of electrons can form *ionic bonds*. For instance, the calcium atom has an electron configuration of 2 electrons in its outer shell. The chlorine atom has 7 outer electrons. If the calcium atom transfers 2 electrons, one to each chlorine atom, it becomes a calcium ion with the stable configuration of an inert gas. At the same time, each chlorine, having gained 1 electron, becomes a chlorine ion, also with an inter-gas configuration. The bonding in calcium chlorine is the electrostatic attraction between the ions.

Covalent bonds are formed by sharing of *valence* (the number of electrons an atom can give up or acquire to achieve a filled outer shell) electrons. Hydrogen atoms, for instance, have 1 outer electron. In the hydrogen molecule, H_2, each atom contributes 1 electron to the bond. Consequently, each hydrogen atom has control of 2 electrons— one of its own and the second from the other atom—giving it the electron configuration of an inert gas. In the water molecule, H_2O, the oxygen atom, with 6 outer electrons, gains control of 2 extra electrons supplied by the 2 hydrogen atoms. Similarly, each hydrogen atom gains control of an extra electron from the oxygen.

Chemical compounds are often classified into either of 2 groups based on the nature of the bonding between their atoms. And, as you might expect, chemical compounds consisting of atoms bonded together by means of ionic bonds are called *ionic compounds*. Compounds whose atoms are bonded together by covalent bonds are called *covalent compounds*.

Some interesting contrasts are possible between most ionic and covalent compounds. For example, ionic compounds have higher melting points, boiling points, and solubility in water than covalent compounds. Ionic compounds are nonflammable, while covalent compounds are flammable. Ionic compounds that are molten in water solutions conduct electricity. Molten covalent compounds do not conduct electricity. Ionic compounds generally exist as solids at room temperature, while covalent compounds exist as gases, liquids, and solids at room temperature.

Water Chemistry

A colorless, odorless, tasteless liquid, water is the only common substance that occurs naturally on the earth in all 3 physical states: solid, liquid, and gas. Approximately 73% of the earth's surface, almost 328 million cubic miles, is covered with water. The human body is 70% water by weight, and water is essential to the life of every living thing.

Water is a simple, unique molecule. It is composed of 1 oxygen atom and 2 hydrogen atoms. Water is a "polar" molecule, meaning that there is an uneven distribution of electron density. One side of the molecule is positive, the other is negative. Water molecules orient to neutralize electric charges; this buffering action keeps foreign ions from reacting and precipitating. Pure water itself resists ionizing, and is a poor electrical conductor. These properties give water an amazing capacity to dissolve other molecules. It is, in fact, considered the universal solvent.

Many of the unique properties of water are due to hydrogen bonding. For example, the unique physical properties, including a high heat of vaporization, strong surface tension, high specific heat, and the aforementioned nearly solvent properties of water are also due to hydrogen bonding. Hydrogen bonding also causes another unique property called the *hydrophobic effect* (water hating, repelling; incapable of dissolving in water), or the exclusion of compounds containing carbon and hydrogen (nonpolar compounds). The hydrophobic effect is particularly important in the formation of cell membranes. Basically, we can say that water "squeezes" nonpolar molecules together.

✓   **Important Point**: Hydrophobic substances (e.g., nonpolar compounds) do not dissolve in water. Hydrophilic substances do dissolve in water.

✓   **Interesting Point**: Typically, when we think about the heating or cooling of a substance, we think of the substance expanding with the heat and contracting when it is cooled. Fortunately for life as we know it, this is not the case with water. Because of hydrogen bonding between water molecules, it does just the opposite. The result is that ice is less dense than water. Ice floats on top of the surface, releasing heat from the water below and acting as an insulator. If this were not the case, bodies of water would freeze from the bottom up, making life impossible for almost anything living in them.

Acids, Bases, pH

When acids and bases are combined in the proper proportions they neutralize each other (they are chemical opposites), each losing its characteristic properties and forming a salt and water.

$$Acid + Base = Salts$$

The acid-base-salt concept originated with the beginning of chemistry and is very important in industrial chemicals, in the environment, and life processes.

The word *acid* is derived from the Latin *acidus*, which means sour. The sour taste is one of the properties of acids (however, you should never actually taste an acid in the lab or anywhere else). An acid is a substance that, in water, produces hydrogen ions, H^+, and has the following properties:

1. Conducts electricity
2. Tastes sour
3. Changes the color of blue litmus paper to red
4. Reacts with a base to neutralize its properties
5. Reacts with metals to liberate hydrogen gas

A *base* is a substance that produces hydroxide ions, OH⁻, and/or accepts H⁺, and when dissolved in water, has the following properties:

1. Conducts electricity
2. Changes the color of red litmus paper to blue
3. Tastes bitter and feels slippery
4. Reacts with an acid to neutralize the acid's properties

A common way to determine whether a solution is an acid or a base is to measure the concentration of hydrogen ions (H⁺) in the solution. The concentration can be expressed in powers of 10, but is more conveniently expressed as *pH* (potential of hydrogen). That is, pH is a logarithmic measure of hydrogen ion concentration, originally defined by Danish biochemist S. P. L. Sorensen in 1909.

$$\text{pH} = -\log[\text{H}^+]$$

For example, pure water has 1×10^{-7} grams of hydrogen ions per liter. The negative exponent of the hydrogen ion concentration is called the pH of the solution. The pH of water is 7; a neutral solution. A concentration of 1×10^{-12} has a pH of 12.

The *pH scale* (see Table 2.4) is used to assess the degree of acidity or basicity (alkalinity) in a solution using a scale ranging from 0 to 14. A pH of 7.0 is considered neutral. Below pH 7 is acidic and above pH 7 is basic.

Table 2.4.   Approximate pH of Some Common Substances

Substance	pH
Battery acid	0.0
Gastric juice	1.2
Lemons	2.3
Vinegar	2.8
Soft drinks	3.0
Apples	3.1
Grapefruit	3.1
Wines	3.2
Oranges	3.5
Tomatoes	4.2
Beer	4.5
Bananas	4.6
Carrots	5.0
Potatoes	5.8
Coffee	6.0
Milk (cow)	6.5
Pure Water ————————Neut———— 7.0————	
Blood (human)	7.4
Eggs	7.8
Sea water	8.5
Milk of magnesia	10.5
Oven cleaner	13.0

✓   **Important Points**: Most biological fluids have a pH range between 6.0 and 8.0. A notable exception is the acidic pH of our stomachs, which has an important role in the digestion of proteins. Keep in mind that the pH scale is logarithmic. Each measure represents a tenfold change. For example, something that is pH 3 is 10 times more acidic than something that is pH 4.

Organic Chemistry

Organic chemistry is the branch of chemistry concerned with compounds of carbon; that is, carbon is the main element of organic compounds. Even though carbon is essential to life as we know it, the study of carbon falls with the realm of inorganic chemistry. Aside from water, most biologically important molecules are carbon-based (organic). Carbons atoms are the most versatile building blocks of molecules. The science of organic chemistry is incredibly complex and varied. Millions of different organic compounds are known today and 100,000+ of these are products of synthesis, unknown in nature.

The molecules of organic compounds have one common feature: one or more carbon atoms that covalently bond to other atoms—that is, pairs of electrons are shared between atoms. Carbon usually forms 4 covalent bonds in compounds due to its 4 valence electrons. The carbon skeletons of organic molecules can vary in length, shape (straight chain, branched, or ring), number and location of double bonds, and elements bonded to available sites. These variations contribute to the diversity of organic molecules. The elements hydrogen, nitrogen, and oxygen are frequently bonded to carbon.

When carbon atoms share electrons with other carbon atoms, 2 carbon atoms may share electrons in such a manner that they form either of the following: carbon-carbon single bonds (C—); carbon-carbon double bonds (C=); carbon-carbon triple bonds (≡). Each bond, written here as a dash (—), is a shared pair of electrons.

FUNCTIONAL GROUPS IN ORGANIC CHEMISTRY

Molecules that have similar properties are classified as *functional groups*, which contribute to the molecular diversity of life. Functional groups are small characteristic groups of atoms that are frequently bonded to the carbon skeleton of organic molecules. These functional groups:

• Behave consistently from one organic molecule to another
• Have specific chemical and physical properties
• Are the regions of organic molecules which are commonly chemically reactive
• Depending upon their number and arrangement, determine unique chemical properties of organic molecules in which they occur

When functional groups are bonded to the carbon atom, the resulting compounds wind up with specific chemical and physical properties (see Table 2.5). Diverse organic

Table 2.5.   Functional Groups Important in Biology

Functional Group	Comments
Amino Group ($-NH_2$)	Consists of a nitrogen atom bonded to 2 hydrogens and to the carbon skeleton. Is an important functional group in amino acids, and, hence, proteins. It is polar and, therefore, soluble in water. Organic compounds with this functional group are called amines. It acts as a weak base.
Aldehydes	Has C=O at end of carbon chain; and ketones have C=O in a place other than the end of a carbon chain—both are important in sugars.
Sulfhydryl Group ($-SH$)	Group consists of an atom of sulfur bonded to an atom of hydrogen; they help stabilize the structure of proteins that have sulfur in their structure. Organic compounds containing sulfhydryls are called thiols.
Phosphate Group	Group composed of phosphoric acid (H3PO4) and carbon skeleton. This group is extremely polar. Organic phosphates are very important in the cellular storage and transfer of adenosine triphosphate (ATP), a compound that releases free energy when its phosphate bonds are hydrolyzed.
Methyl Group ($-CH_3$)	Group is nonpolar and hydrophobic. It contributes to one of the principal structures of proteins.

molecules found in living organisms have carbon skeletons. These molecules can be viewed as hydrocarbon derivatives with functional groups in place of H, bonded to carbon at various sites along the molecule.

✓ **Key Point**: Diverse organic molecules found in living organisms have carbon skeletons.

Macromolecules

Macromolecules are very large, long molecules consisting of many smaller structural units linked together. Biochemistry, the study of the biological chemistry of life, is the study of the structure, function and interactions of the 4 types of macromolecules: carbohydrates, lipids, proteins, and nucleic acids. These are the basic building blocks used to construct cells, the basic operational units from which the great diversity of life has evolved.

✓ **Important Point**: Macromolecules are polymers (chains) of repeating links called monomers.

• **Carbohydrates**—The generalized chemical formula for carbohydrate is CH_2O. Carbohydrates are the major caloric intake for most animals. They are used by all organisms as a quick energy source, to do work, when carbohydrates are broken down.

Carbohydrates are also the structural material in plants (trees). Carbohydrate sugars occur naturally in fruits, milk, milk products, honey, corn syrup, and some vegetables. Carbohydrates include monosaccharides, disaccharides, and polysaccharides. *Monosaccharides* are simple sugars. Examples of monosaccharides include glucose (most common sugar in human body), galactose (in milk), fructose (honey), ribose and deoxyribose. *Disaccharides* are formed by linking 2 monosaccharide molecules, as reservoirs of energy. Sucrose, a common plant disaccharide is composed of the monosaccharides glucose and fructose. Other examples of disaccharides include maltose and lactose. Disaccharides are used by many organisms as reservoirs of energy. *Polysaccharides* are insoluble large molecules composed of individual monosaccharide units linked by dehydration. Polysaccharides formed from glucose are starches and are important plant products, for example, potato starch. Animals construct the insoluble polysaccharide *glycogen* as their principal form of glucose. Glycogen is an animal storage product that accumulates in the vertebrate liver. The polysaccharide cellulose is the main structural constituent of plant cell walls and is very difficult to break down. Cellulose forms the fibrous part of the plant cell wall. When insects and fungi add a nitrogen group to the glucose units of cellulose, *chitin*, a highly resistant structural material found in anything from the shells of beetles to webs of spiders, is produced.

- **Lipids**—There are many different kinds of *lipids*, which are nonpolar (hydrophobic) and are, therefore, insoluble in water. Lipids are an essential part of a balanced diet, playing a critical role in the body where they are involved mainly with long-term energy storage. Lipids that are solid at room temperature are fats, or oil if liquid. Approximately 95% of the lipids in food and human body are fats. Along with providing long-term energy storage, lipids (fat) also provide body insulation, shock absorbing (cushioning) qualities, lubricants, and oils. The human body naturally accumulates some fats in the posterior area. Under the skin (subdermal) fat plays a role in insulation. Lipids may be triglycerides, or mono- and diglycerides. They are saturated or unsaturated. Other important lipids include phospholipids (important structural compounds of cell membranes), steroids such as cholesterol and the sex hormones, and prostaglandins, modified fatty acids which act as local chemical messengers in vertebrates.

- **Proteins**—Proteins, composed of natural polymer molecules consisting of amino acids, are probably the most important class of biochemical molecules, although of course lipids and carbohydrates are also essential for life. Proteins are the basic constituents in all living organisms and are the basis for the major structural compounds of animal and human tissue. Control functions of proteins are carried out by enzymes and proteinaceous hormones. Enzymes are chemicals that act as organic catalysts (a catalyst is a chemical that promotes but is not changed by a chemical reaction). Proteins have a wide range of important and varied functions, which include:

  - Serving as enzymatic catalysts
  - Use as transport molecules
  - Use as storage molecules
  - Use in movement of muscles

- Use in mechanical support
- Mediating cell response
- Controlling cell growth and differentiation
- Providing defense mechanisms against foreign substances and disease-causing organisms
- Signaling

- **Nucleic Acids**—Nucleic acid is a complex, high-molecular-weight biochemical macromolecule composed of monomer units known as nucleotide chains that convey genetic information. Primarily, nucleic acids are biology's means of storing and transmitting genetic information, though RNA is also capable of acting as an enzyme. Along with the function of information storage (DNA, a double-stranded helix that contains the sugar deoxyribose), the main function of nucleotides is protein synthesis (RNA—ribonucleic acid, a single-stranded molecule, contains the sugar, ribose) and energy transfers (ATP and NAD).

Chapter Review Questions

1. List 4 chemical reactions that occur all around us.
2. What is matter?
3. Define *element*.
4. What is a compound?
5. They make up about 75% of the elements on the periodic table: _____.
6. They have properties resembling both metals and nonmetals: _____.
7. Define *physical change*.
8. The _____ of an element is equal to the total number of protons and neutrons in the nucleus of an atom of an element.
9. What is a chemical bond?
10. Acid + Base = _____.
11. _____ is the main element of organic compounds.
12. What are functional groups?
13. What are the 4 macromolecules?
14. The structural materials of plants: _____.
15. Fats are made up of macromolecules known as: _____.

References and Additional Reading

American Heritage Dictionary, 4th ed., s.v. "Chemistry." Accessed July 11, 2006, at http://education.yahoo.com/reference/dictionary/.

Manahan, S., 2004. *Environmental Chemistry*, 8th ed. New York: Lewis Publishers.

Mathews, C. K., and K. E. van Holde.1996. *Biochemistry*, 2nd ed. Menlo Park, CA: Benjamin/ Cummings.

Raven, P. H., and G. B. Johnson.1992. *Biology*, 3rd ed. Missouri: Mosby-Year Book, Inc.

Research Machine. 2006. *Chemistry*. Accessed June 28, 2006, at www.tiscali.co.uk/reference/ encyclopaedia/Hutchinson/m0000617.html.

Spellman, F. R., and N. E. Whiting. 2006. *Environmental Science and Technology*, 2nd ed. Rockville, MD: Government Institutes.

Thoreau, H.D. 1994. *Walking (1862)*. San Francisco: Harper.

CHAPTER 3

Cells: The Fundamental Units of Life

Amoebas at the start
Were not complex;
They tore themselves apart
And started Sex.

—Arthur Guiterman (1871–1943)

Topics in This Chapter

- Types of Cells
- Cell Structure
- Intercellular Junctions (Animal Cells)

Cells are the fundamental units of life. The cell retains a dual existence as distinct entity and a building block in the construction of organisms. These conclusions about cells were observed and published by Schleiden (1838). Later, Rudolph Virchow added the powerful dictum, "Omnis cellula e cellula," in other words, "All cells only arise from pre-existing cells." This important tenet, along with others, formed the basis of what we call *cell theory*. The modern tenets of the cell theory include:

- All known living things are made up of cells.
- The cell is the structural and functional unit of all living things.
- All cells come from pre-existing cells by division.
- Cells contain hereditary information which is passed from cell to cell during cell division.
- All cells are basically the same in chemical composition.
- All energy flow of life occurs within cells.

The modern tenets, of course, post-dated Robert Hooke's 1663 discovery of cells in a piece of cork, which he examined under his primitive microscope. Hooke drew the cell (actually it was the cell wall he observed) and coined the word *cell*. The word *cell* is derived form the Latin word *cellula*, which means small compartment.

Thus, since the 19th century, we have known that all living things, whether animal or plant, are made up of cells. Again, the fundamental unit of all living things, no matter how complex, is the cell. A typical cell is an entity isolated from other cells by a membrane or cell wall. The cell membrane contains protoplasm (the living material found within them) and the nucleus.

In a typical mature plant cell (see Figure 3.1), the cell wall is rigid and is composed of nonliving material, while in the typical animal cell (see Figure 3.2) the wall is an elastic living membrane. Cells exist in a very great variety of sizes and shapes, as well as functions. The cell is the smallest functioning unit of a living thing that still has the characteristics of the whole organism. Sizes range from bacteria too small to be seen

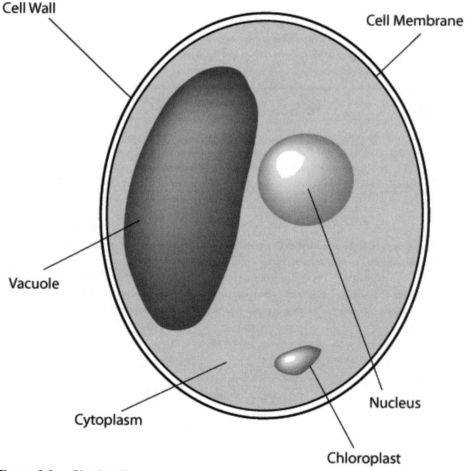

Figure 3.1.   Plant cell.

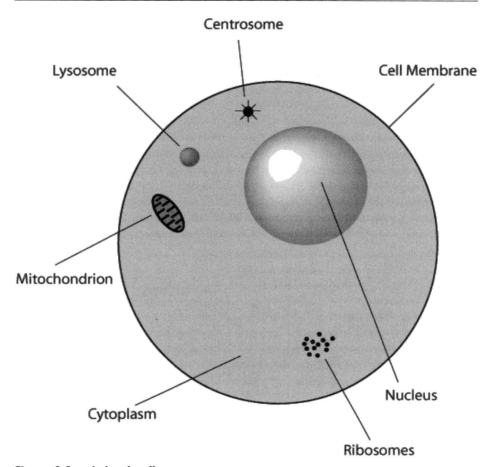

Figure 3.2.   Animal cell.

with the light microscope to the largest single cell known, the ostrich egg. Microbial cells also have an extensive size range, some being larger than human cells.

✓   **Interesting Point**: How small a cell can be is limited by the minimum volume required to contain the genetic material, proteins, etc., which are necessary to carry out basic cell functions and reproduction. How large a cell can be is limited by metabolism. A cell must take in adequate amounts of oxygen and nutrients and get rid of wastes.

Types of Cells

Cells are of 2 fundamental types, prokaryotic and eukaryotic. ***Prokaryotic*** (meaning "before nucleus") cells are simpler in design than eukaryotic cells, possessing neither a nucleus nor the organelles (i.e., internal cell structures, each of which has a specific function within the cell) found in the cytoplasm of *eukaryotic* (meaning "true nucleus") cells. Because prokaryotes do not have a nucleus, DNA is in a "nucleiod" region. With the exception of archaebacteria, proteins are not associated with bacterial DNA. Bacteria are the best known and most studied form of prokaryotic organisms (see Figure 3.3).

✓   **Important Point**: Cells may exist as independent units (e.g., the protozoa), or as parts of multicellular organisms in which the cells may develop specializations and form tissues and organs with specific purposes.

Prokaryotes are unicellular organisms that do not develop or differentiate into multicellular forms. Some bacteria grow in filaments, or masses of cells, but each cell in the colony is identical and capable of independent existence. Prokaryotes are capa-

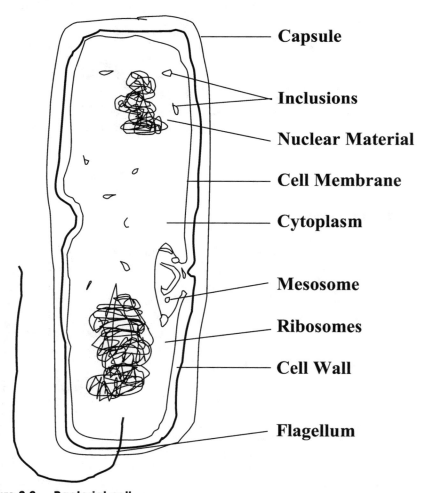

Figure 3.3.   Bacterial cell.

ble of inhabiting almost every place on the earth, from the deep ocean, to the edges of hot springs, to just about every surface of our bodies.

✓   **Important Point**: It is often stated that prokaryotic cells are among the most primitive forms of life on Earth. However, it is important to point out that primitive does not mean they are outdated in the evolutionary sense, since primitive bacteria seem little changed, and thus may be viewed as well adapted.

As mentioned, prokaryotes are distinguished from eukaryotes on the basis of nuclear organization, specifically their lack of a nuclear membrane. Also, prokaryotes are smaller and simpler than eukaryotic cells. Again, prokaryotes also lack any of the intracellular organelles and structures that are characteristic of eukaryotic cells. Most of the functions of organelles, such as mitochondria, chloroplasts, and the Golgi apparatus, are taken over by the prokaryotic plasma membrane. Prokaryotic cells have 3 architectural regions: appendages called **flagella** and **pili**—proteins attached to the cell surface; a **cell envelope** consisting of a **capsule**, a **cell wall**, and a **plasma membrane**; and a **cytoplasmic region** that contains the **cell genome** (DNA), ribosomes, and various sorts of inclusions.

Eukaryotic cells evolved about 1.5 billion years ago. Protists, fungi, plants, and animals have eukaryotic cells—all plants and animals are eukaryotes. They are larger, as much as 10 times the size of prokaryotic cells, and most of their genetic material is found within a membrane-bound nucleus (a true nucleus), which is generally surrounded by several membrane-bound organelles. The presence of these membrane-bound organelles points to the significant difference between prokaryotes and eukaryotes. Although eukaryotes use the same genetic code and metabolic processes as prokaryotes, their higher level of organizational complexity has permitted the development of truly multicellular organisms.

✓   **Important Point**: An enormous gap exists between prokaryote cells and eukaryote type cells:

> . . . prokaryotes and eukaryotes are profoundly different from each other and clearly represent a marked dichotomy in the evolution of life . . . The organizational complexity of the eukaryotes is so much greater than that of prokaryotes that it is difficult to visualize how a eukaryote could have arisen from any known prokaryote. (Hickman et al. 1997)

Prokaryotic and eukaryotic cells also have their similarities. All cell types are bounded by a plasma membrane that encloses proteins and usually nucleic acids such as DNA and RNA. Table 3.1 shows a comparison of key features of both cell types.

✓   **Interesting Point**: Plant cells can generally be distinguished from animal cells by: (1) the *presence* of rigid cell walls (composed of nonliving matter), chloroplasts, and central vacuoles in *plants* and their absence in animals, and (2) the *presence* of a cell membrane, lysosomes and centrioles in *animals* and their absence in plants (Spellman 2000).

Table 3.1.   Comparison of Typical Prokaryotic and Eukaryotic Cells

Characteristic	Prokaryotic	Eukaryotic
Size	1–10 μm	10–100 μm
Nuclear envelope	Absent	Present
Cell wall	Usually	Present (plants)/Absent (animals)
Plasma membrane	Present	Present
Nucleolus	Absent	Present
DNA	Present (single loop)	Present
Mitochondria	Absent	Present
Chloroplasts	Absent	Present (plants only)
Endoplasmic reticulum	Absent	Present
Ribosomes	Present	Present
Vacuoles	Absent	Present
Golgi apparatus	Absent	Present
Lysosomes	Absent	Often Present
Cytoskeleton	Absent	Present

Cell Structure

As mentioned, cells contain structures called organelles. The structure and function of the major organelles in eukaryotic and prokaryotic cells are described below.

- **Plasma membrane**—is located between the cell and its environment; it is a semi-permeable phospholipid bilayer. The membrane serves to separate and protect a cell from its surrounding environment and is made mostly from a double layer of proteins and lipids, fat-like molecules. The proteins within the membrane have a variety of functions. For example, some proteins are receptors which can detect the presence of certain kinds of molecules in the surrounding fluids. Also embedded within this membrane are a variety of other molecules that act as channels and pumps, moving different molecules into and out of the cell. Cells that are specialized for absorption (e.g., intestinal cells) have folds in the plasma membrane called *microvilli* that increase the surface area. Pseudopodia are temporary extensions of the plasma membrane used for movement or to engulf particles. A form of plasma membrane is also found in prokaryotes, but is usually referred to as the cell membrane.
- **Nucleus (cell's control center)**—is the control center of the cell. The nucleus contains DNA and is the place where RNA synthesis occurs; it is therefore the cell's control center because DNA contains instructions needed to produce proteins that control metabolism and other cell functions. **Chromatin** is the grainy threadlike DNA. During cell division, the nuclear membrane disintegrates and the DNA becomes coiled, producing visible structures called **chromosomes**. The material within the nucleus is referred to as the nucleoplasm. The nucleus is spheroid in shape and separated from the cytoplasm by a double membrane called the **nuclear envelope**. The nuclear envelope isolates and protects a cell's DNA from various molecules that could accidentally damage its structure or interfere with its processing. **Nuclear pores** allow materials to pass into and out of the nucleus. During processing, DNA is transcribed, or synthesized, into a special RNA, called mRNA. This mRNA is

then transported out the nucleus, where it is translated into a specific protein molecule. In prokaryotes, DNA processing takes place in the cytoplasm.

- **Nucleolus**—is a structure made up of RNA and proteins within the nucleus where the ribosomal subunits are produced. The ribosomes exit pores in the nuclear envelope, and enter the cytoplasm, where they are involved in protein synthesis.

- **Endoplasmic reticulum (ER)**—within the eukaryotic cell: a system of membranes that ramifies through the cytoplasmic region and forms the limiting boundaries, compartments, and channels whose lumina are completely isolated from the cytoplasm; the ER is a protein-containing lipid bilayer. The ER is the transport network for molecules targeted for certain modifications and specific destinations, as compared to molecules that will float freely in the cytoplasm. ER has 2 forms that differ in structure and function: the **rough ER** and the **smooth ER**. In the *rough ER*, the rough appearance is due to the presence of ribosomes on the membrane. The rough ER functions in protein synthesis, especially "proteins that are to be secreted to outside the cell. Proteins enter the interior (lumen) of the endoplasmic reticulum while being synthesized. The rough ER also functions in the modifications of newly formed proteins. For example, some enzymes may add carbohydrate chains forming glycoproteins. Other enzymes function to hold the newly synthesized proteins into their proper shape" (Clinton Community College 2006). In regard to transport of molecules through the rough ER, the molecules go through the internal channel of the ER and fold into their own conformation. The ER surrounds the molecules with "vesicles (sacs) that pinch off the ER or Golgi apparatus (discussed below) and transport the molecules (e.g., secretory proteins such as insulin) to other parts of the cell" (Spellman & Whiting 1999; Clinton Community College 2006).

"The *smooth ER* has no ribosomes attached to it. As mentioned, it is continuous with rough ER. The smooth ER reticula have a variety of different functions but often function to produce lipid compounds such as phospholipids, steroids, and fatty acids. Certain kinds of cells have smooth endoplasmic reticulum with a specialized function. For example, smooth ER in the adrenal cortex and testes produces steroid hormones; the smooth ER of liver cells helps detoxify drugs in the blood; and calcium ions needed for contraction are stored in the smooth ER of muscle cells. Vesicles pinch off the smooth ER and carry materials to other parts of the cell such as the Golgi apparatus or the plasma membrane" (Spellman & Whiting 1999; Clinton Community College 2006).

- **Golgi apparatus** (sometimes called a Golgi body or Golgi complex)—is a stack 3 to 20 flattened, slightly curved sacs which appear like a stack of pancakes that is a continuation of the ER. The Golgi apparatus is where exportable vesicles of ER proteins are passed into the forming face for further processing, modifying, and packaging, and transported from the maturing face to a variety of other cellular locations.

- **Lysosomes** ("bodies that dissolve")—are formed in the Golgi apparatus and are often referred to as the garbage disposal system of the cell. Lysosomes are somewhat spherical, membrane-bounded vesicles of hydrolytic enzymes, enabling them to digest all major classes of macromolecules. Lysosomes can contain more than three dozen enzymes for degrading proteins, nucleic acids, and certain sugars called polysaccharides. Lysosomes also digest foreign bacteria that invade the cell. All of these enzymes work best at an optimal acidic environment (pH 5.0). Lysosomes point out the importance behind compartmentalization of the eukaryotic cell. The cell could not house such destructive enzymes if they were not contained in a membrane-bound system.

- **Mitochondria**—are commonly referred to as the powerhouse of the cell. Mitochondria are self-replicating organelles that contain their own DNA and occur in various numbers, shapes and sizes in the cytoplasm of all eukaryotic cells. Mitochondria have an external membrane and an inner membrane with numerous folds called *cristae*. Cristae project into the gel-like matrix. Enzymes involved in cellular respiration are found in the matrix and embedded in the membrane of the cristae. Cristae provide more surface area for chemical reactions to occur within mitochondria. Mitochondria are composed of 3 types of protein fibers: microtubules, actin filaments, and intermediate filaments. These protein fibers function to move materials within the cell, move the cell, and provide mechanical support. Mitochondria convert fuel to usable energy. They are semiautonomous and self-replicating.

- **Peroxisomes**—are vesicles found in nearly all eukaryotes that function to rid the body of toxic substances, such as hydrogen peroxide, or other metabolites, and contain enzymes concerned with oxygen utilization. Peroxisomes often resemble a lysosome. However, peroxisomes are self-replicating, whereas lysosomes are formed in the Golgi apparatus. Peroxisomes also have membrane proteins that are critical for various functions, such as for importing proteins into their interiors and to proliferate and segregate into daughter cells. Peroxisome enzymes are also involved in the breakdown of fatty acids to acetyl CoA, which is transported to the mitochondria for fuel during cellular respiration. The liver, where toxic byproducts accumulate, contains high numbers of peroxisomes.

- **Vacuoles**—are membranous sacs similar to but larger than vesicles. They store water and dissolved substances. They are more important in plant cells. Most of the center of a plant cell is occupied by a long *central vacuole*, which gives support because pressure within the vacuole makes the cell turgid (rigid). The cell wall prevents the cell from bursting. Some organisms (e.g., some fresh-water protozoa) have specialized *contractile vacuoles* for eliminating excess water and *food vacuoles* (formed by phagocytosis, the process in which white blood cells consume particles) that contain food within he cell.

- **Ribosomes**—are cytoplasmic granules composed of RNA and protein, where protein synthesis takes place. Ribosomes are large complexes composed of many molecules found in both prokaryotes and eukaryotes. Protein synthesis is the process by which proteins are made from individual amino acids. Ribosomes read the code in messenger RNA (mRNA) and synthesize protein accordingly. Several ribosomes may be attached to a strand of mRNA forming a unit called a *polysome*. The process of converting mRNA's genetic code into the exact sequence of amino acids that make up a protein is called *translation*. A ribosome is composed of two subunits, one large and one small, each having a different function during protein synthesis. In eukaryotic cells, the subunits are synthesized in the *nucleolus* and move into the cytoplasm. During the process of protein synthesis, 2 subunits will come together along with mRNA. Ribosomes are composed of both RNA (called ribosomal RNA and tRNA) and protein. Ribosomes in eukaryotes are about 33% larger than those in prokaryotes.

- **Cell wall**—functions to give shape, support and protection to the cell. Cell walls are found in almost all plant cells. The cell wall surrounds the cell on the outside and is the secretory product of the cytoplasm. It is permeable to water. Plants have cell walls composed of cellulose; fungi have walls composed of chitin.
- **Cytoskeleton** (the cell's scaffold)—is an important, complex, and dynamic cell component. It acts to organize and maintain the cell's shape, anchors organelles in place, helps with the cell's uptake of external materials during endocytosis, and moves parts of the cell in processes of growth and motility. There are a great number of proteins associated with the cytoskeleton, each controlling a cell's structure by directing, bundling, and aligning filaments. There are 3 types of cytoskeleton filaments or fibers: microtubules, microfilaments, and intermediate filaments.

  - **Microtubules**—are in all eukaryotic cells. They are straight, hollow fibers made from globular proteins called *tubulin*. They have many functions including cellular support, organelle movement, cell motility (they move the cilia and flagella), and the separation of chromosomes during cell division. The assembly of microtubules is controlled by an area near the nucleus called the *centrosome*, or microtubule organizing area.
  - **Microfilaments**—are made up of 2 intertwined strands of the protein, actin. They provide cellular support when they combine with other proteins just inside of the plasma membrane and play a role in cell shape. They participate in muscle contraction and localized contraction of cells (i.e., they move the cell).
  - **Intermediate filaments**—are composed of keratin subunits and are more permanent than either microtubules or microfilaments. They may be the framework of the cytoskeleton; they provide mechanical support. They reinforce cell shape and probably fix an organelle's position in the cell. These fibers line the interior of the nuclear envelope.

- **Cytoplasm** (cell's inner space)—is the material that lies within the cytoplasmic membrane, or the membrane that surrounds a cell. It contains none of a cell's genetic material, because this is contained in the nucleus. It does, however, contain a lot of water, and the other organelles of the cells. It provides a platform upon which they can operate within the cell. It is made up of proteins, vitamins, ions, nucleic acids, amino acids, sugars, carbohydrates, and fatty acids. All of the functions for cell expansion, growth, and replication are carried out in the cytoplasm of a cell. The cytoplasm also contains many salts and is an excellent conductor of electricity, creating the perfect environment for the mechanics of the cell. The function of the cytoplasm, and the organelles which reside in it, are critical to a cell's survival.
- **Chloroplasts**—are large, complex double-membraned organelles that perform the function of photosynthesis within plant cells and contain the substance chlorophyll that is essential for the process. The chloroplasts use photosynthetic chlorophyll pigment and take in sunlight, water and carbon dioxide to produce glucose (sugar) and oxygen. This is the process of photosynthesis. Chloroplasts make all the food for other organelles.
- **Cilia and flagella**—are hairlike structures projecting from the cell that function to move the cell by their movements (e.g., sperm are motile; they use flagella to move).

They contain cytoplasm and are enclosed by the plasma membrane. Cells that contain cilia are *ciliated*. Cilia are shorter than flagella but are similar in construction. Cilia and flagella are formed from a core of 9 outer microtubules and 2 inner single microtubules ensheathed in an extension of the plasma membrane.

Intercellular Junctions (Animal Cells)

The way cells interact with neighbor cells varies, but they are all virtually in contact with each other. Contact is maintained or effected via *cell junctions*. These junctions anchor cells to one another to provide a passageway for cellular exchange. There are at least 3 different types of cell junctions (contacts): desmosomes, tight junctions, and gap junctions.

- **Desmosomes**—are strong protein attachments between adjacent animal cells. They are interacting complementary folds of membrane. Desmosomes act like spot welds or interlinking fingers to hold together tissues, such as skin or heart muscle tissues that undergo considerable stress.
- **Tight junctions**—are tight bands of proteins that prevent fluids and small molecules from crossing the membrane. The junction completely encircles each cell, preventing the movement of material between the cells. Tight junctions in the stomach lining protect the stomach cells from hydrochloric acid and are characteristic of animal cells lining the digestive tract, where materials are required to pass through cells to penetrate the blood stream.
- **Gap junctions**—are narrow tunnels (doorways) between animal cells that consist of proteins called *connexions*. The proteins prevent the cytoplasms of each cell from mixing, but allow the passage of ions and small molecules. In this manner, they allow the flow of materials and electrical charge.

Chapter Review Questions

1. The plasma membrane consists primarily of:
2. Microtubules are characterized by:
3. Lysosomes are involved in the _____ of cellular substances.
4. _____ is/are often more numerous near areas of major cellular substances.
5. Plant and animal cells differ mostly in that only plant cells have _____.
6. A smooth ER manufactures _____.
7. Components of the cell wall include:
8. The 2 categories of cells are:
9. Organelle containing the pigment chlorophyll:

THOUGHT-PROVOKING QUESTION

1. Describe the structure of the plasma membrane and the various ways in which it permits interactions with the outside environment.

References and Additional Reading

Cibas, E.S., and B.S. Ducatman. 2003. *Cytology: Diagnostic Principles and Clinical Correlates*. London: Saunders Ltd.

Clinton Community College, 2006. *Welcome to the Biology Web: Cells*. Accessed June 22, 2006, at http://faculty.clintoncc.suny.edu/faculty/Michael.Gregory/files/Bio%20100/Bio%20100%20Lectures/Cells/cells.htm.

Creekmore, T. 2006. *The Science Channel: 100 Greatest Discoveries: Biology*. Atlanta, GA: Discovery Communication.

DeDuve, C. 1984. *A Guided Tour of the Living Cell*. New York: Scientific Library. W.H. Freeman Company.

Finean, J.B. 1984. *Membranes and Their Cellular Functions*. Oxford, Boston: Blackwell Scientific Publications.

Frank, J., et al. 1995. A model of synthesis based on cryo-electron microscopy of the E. coli ribosome. *Nature* 376: 440–444.

Garrett, R., et al. 2000. *The Ribosome: Structure, Function, Antibiotics, and Cellular Interactions*. Washington, DC: American Society Microbiology.

Hickman, C.P., et al. 1997. *The Biology of Animals*, 7th ed. New York: William C. Brown/McGraw Hill.

Martin, S. 1981. *Understanding Cell Structure*. Cambridge, New York: Cambridge University Press.

Murray, A.W. 1993. *The Cell Cycle: An Introduction*. New York: W.H. Freeman.

Serafini, A. 1993. *The Epic History of Biology*. New York: Plenum.

Spellman, F.R. (2000). *Microbiology for water/wastewater operators*, rev. ed. Lancaster, PA: Technomic Publishing Company.

Spellman, F.R., and N.E. Whiting, 1999. *Environmental Science and Technology: Concepts and Applications*. Rockville, MD: Government Institutes.

Spirin, A. 1986. *Ribosomes Structure and Protein Biosynthesis*. CA: The Benjamin/Ammins Publishing Co. Inc.

Thomas, L. 1995. *The Life of a Cell: Notes of a Biology Watcher*. New York: Penguin Books.

CHAPTER 4

Energy

The uniformity of earth's life, more astonishing than its diversity, is accountable by the high probability that we derived, originally, fromsome single cell, fertilized in a bolt of lightning as the earth cooled.

—Lewis Thomas (1978, 5)

Topics in This Chapter

- Laws of Thermodynamics
- ATP
- Enzymes
- Cellular Respiration and Metabolism
- Photosynthesis

It cannot be seen, tasted, smelled, or touched. What is it? Well, maybe a quote from Alice in *Through the Looking Glass* is appropriate here: "It seems very pretty, but it's rather hard to understand." In this case, we are referring to energy. Definitely difficult for many of us to understand, this is not to say, of course, that we do not occasionally think about energy. However, it is more often the case that we take energy for granted through a deceptive familiarity, because we think of it in so many different ways: electrical energy, atomic energy, chemical energy (e.g., batteries), heat energy, radiant energy, sound energy, ultraviolet (UV) energy, cheap energy, expensive energy (have you checked out the recent average cost of a gallon of gasoline—$3-plus and apparently on the rise, with no terminal price-end in sight?), food energy, abundant energy, kinetic energy, potential, and so on.

Again, for many of us, energy is difficult to understand. However, like many other things in life, we may not be able to explain it but we certainly are aware of it. For example, when something gets hotter or colder, energy is involved. When something speeds up or slows down, energy is involved. When something gives off light, energy is involved. When 2 chemicals react, energy is involved. When plants grow, energy is involved.

From the preceding list, it should be apparent that for energy to be useful (i.e., for what we perceive as being useful) it must be transformed from one form to another. Simply put, to be useful energy must be transformed. This is certainly the case in processes essential to life.

"There are a number of energy transformations in plants and animals that are essential to life. For example, both plants and animals use energy to make the complex molecules necessary for life. One example is the production of DNA" (Nave 2007), another is the process of growth, which requires a lot of energy to create the new cells and enlarge the structures. Another form of transformed energy is electricity. Consider, for example, the following energy transformation:

$$wind \rightarrow mechanical \rightarrow electrical$$

We all know about electricity that lights the light bulb, powers our air conditioners, cools our food, starts and maintains our automobile engines—electrically operated devices or machines. However, it is important to point out that we should also include ourselves as "electrically operated machines, because we are. Each of our cells has an electric voltage or potential associated with it. This voltage helps to control the migration of ions across the cell membranes. Probably the foremost example of electrical work is in the operation of the human nerves. When nerves fire, they generate an electrical impulse called an *action potential* which communicates information to the brain, or carry a signal from the brain to a muscle to initiate its movement. Moreover, electrical energy transformation is essential for sensing our environment as well as for reacting to that environment in any way" (Nave 2007).

Probably the most easily recognized use of energy is the mechanical energy used to move muscles, which in turn, move limbs. The muscle movement is very important and requires a lot of energy generated from ATP (discussed later in this chapter). The foundation of living energy begins at the cellular level. In cells, both kinetic and potential energy are present. Kinetic is the energy of pressure and movement, whereas potential is the energy stored up in the cells needed to function and do work. Doing work, in many forms, is what energy is all about.

Laws of Thermodynamics

"The laws of thermodynamics are fundamental truths based on the study of energy exchanges between a system and its surroundings. The *first law of thermodynamics* states that energy cannot be created nor destroyed. In short, energy may change forms in a given process, but we should be able to account for every bit of energy as it takes part in the process. The energy referred to here is that of work, heat, and internal energy. Work is done when a force acts through a distance and the transfer of energy from the system to its surroundings results. Heat is the energy transfer which takes place when 2 regions of different temperatures are connected by a thermal conductor. Internal energy refers to the energy store in a body, and is therefore equal to the heat flow into the system minus the work done by the system on its surroundings" (Eblen & Eblen 1994).

✓   **Important Point**: According to the first law of thermodynamics, the amount of energy in the universe is constant.

The **second law of thermodynamics** states that heat, on its own accord, flows from regions of high temperatures to regions of low temperatures. When applying the law to heat engines, it can be stated that no heat engine can be 100% efficient—no energy conversion is perfect—because some energy is always lost, dissipated as heat (Spellman & Whiting 2006).

When looking at the 2 laws together, it is noted that energy is being constantly degraded; in a thermodynamic process there is always less energy available for doing work, not more. This leads to the definition of **entropy** (every energy conversion increases the disorder and disorganization of matter) which always increases as energy becomes less available for doing work (Eblen & Eblen 1994).

✓   **Important Point**: The second law of thermodynamics has direct application to living cells, because cells cannot transfer or transform energy with 100% efficiency. When we walk, we contract our muscles to produce motion, but heat is always a by-product of this movement.

ATP

"Life, as we know it, is an energy intensive process. It takes energy to make new cells, to extract wastes, heal wounds, operate muscles, and to complete thought processes. All the energy needed is expended in the organism's cells. In some cells, as much as half of the cell's energy output is used to transfer molecules across the cell membrane, a process called *active transport*" (BioMedia Associates 2004).

Again, 50% of the cell's energy output is used to transfer molecules across the cell membrane. Thus, the cells must obtain energy from somewhere. However, neither chemicals from the environment nor sunlight can be used directly to fuel a cell's energy-requiring processes. Therefore, the cell must have ways of **converting** sources of energy into a **usable form** of energy. In the presence of sunlight and certain chemicals, cells can make specific high-energy compounds with which they can satisfy their energy demands; one of these important compounds is adenosine triphosphate (ATP), which is commonly referred to as the "energy currency" of the cell or "universal energy carrier." The energy release from the breakdown of ATP molecules is used to power all bodily functions; it is a source of potential chemical energy for most specialized protein (enzyme) reactions. Proper ATP levels must be maintained for normal cellular function.

✓   **Important Point**: ATP is arguably second in importance only to DNA.

✓   **Interesting Point**: The enormous amount of activity that occurs inside each of the approximately 100 trillion human cells is shown by the fact that at any instant each cell contains about 1 billion ATP molecules. This amount is sufficient for the cell's needs for only a few minutes and must be rapidly recycled. Given 100 trillion cells in the average male, about 1 sextillion (10^{23}) ATP molecules normally exist in the body. "For each ATP the terminal phosphate is added and removed 3 times each minute" (Kornberg 1989).

The process of making ATP involves combining adenosine diphosphate (ADP) and inorganic phosphorus (Pi):

$$ADP + Pi + ENERGY \rightarrow ATP$$

The energy required for this reaction can be obtained in 3 different ways, depending upon the source: photosynthetic phosphorylation (the changing of an organic substance into organic phosphate; photosynthesis is discussed later in the chapter), substrate phosphorylation, or oxidative phosphorylation (occurs on the membranes of mesosomes and related structures of prokaryotes).

✓   **Important Point**: Cells, mostly in the mitochondria, are constantly generating ATP. The chemical energy needed to make ATP comes from food that the cell ingests and digests.

Enzymes

Enzymes are living catalysts (substances that modify and increase the rate of chemical reaction without being consumed in the process) found in every part of the body and involved in every chemical reaction. Essentially proteins, enzymes are formed by the polymerization of some or all the amino acids; 20 amino acids are found in proteins. Enzymes are high molecular weight compounds (ranging from 10,000 to 2,000,000) made up of chains of amino acids linked together by peptide bonds. In the overall linking process, a water molecule is removed between the carboxyl group of one amino acid and the amino group of the next one. Several steps are involved in the actual sequence used for synthesis of proteins, including enzymes, so that chemical energy can be supplied from other molecules.

Most enzymes are pure proteins. However, other enzymes require the participation of small nonprotein groups, which may be organic or inorganic, before their catalytic activity can be exerted. These nonprotein groups are called **cofactors** (the activator). In some cases these cofactors are nonprotein metallic ion activators (ions or iron) that form a functional part of the enzyme. When the cofactor and the protein part (the **apoenzyme**) of the enzyme are present, the entire active complex is called the **holoenzyme**.

Apoenzyme + Cofactor = Holoenzyme

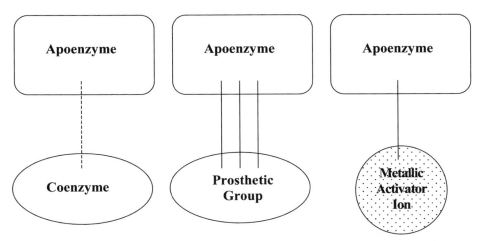

Figure 4.1. Holoenzymes showing apoenzymes and various types of cofactor. Adapted from Witkowski and Power (1975).

The structural nomenclature of enzymes is affected by the way in which the co-factor is attached to the apoenzyme (see Figure 4.1). For example, if the cofactor is firmly attached to the apoenzyme, it is called a **prosthetic** group. When the cofactor is loosely attached to the apoenzyme, it is called a **coenzyme.**

It is important to keep in mind that enzymes increase the speed of reactions, without themselves undergoing any permanent chemical change (they do not alter their equilibrium constants). They are neither used up in the reaction nor do they appear as products of the reaction. This basic enzymatic reaction process is:

Substrate + Enzyme (catalyzes reaction) Product + Enzyme

In the enzymatic reaction process, note that the end product includes the enzyme, which was not altered or destroyed. The enzyme functions by combining in a highly specific way with its **substrate**; the substrate is changed but the enzyme itself is not.

Much research effort has been expended in trying to determine how enzymes lower the activation energy or reactions. What is clear is that enzymes bring substrates together at the enzyme's **active site** to form an **enzyme-substrate complex** (see Figure 4.2). In the enzyme-substrate complex, weak bonds attach the substrate to several points in the active site of the enzyme. This bringing together of enzyme and substrate allows for their concentration, which lowers the activation energy required to complete the reaction. Note that most of these reactions take place at relatively low temperatures ranging from 0° to 36° C.

Enzymes are extremely efficient. Only minute quantities of an enzyme are required to accomplish at low temperatures what normally would require, by ordinary chemical means, high temperatures and powerful reagents. For example, 1 ounce of pepsin can digest almost 2 metric tons of egg whites in a few hours; whereas without

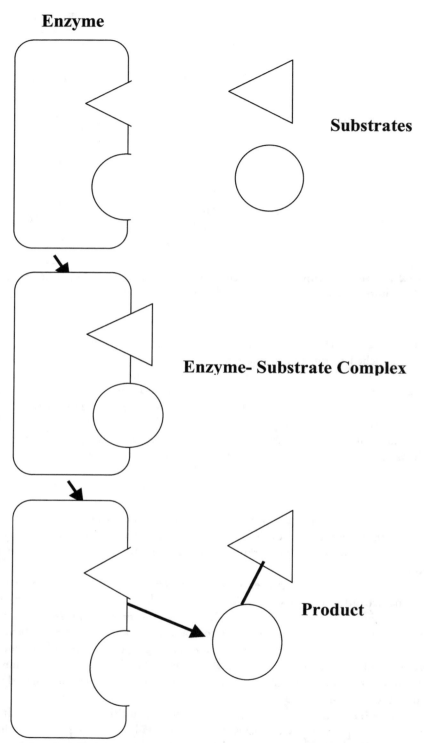

Figure 4.2. Enzyme functions showing the interaction of the substrate and enzyme with the resulting product. Adapted from Prescott et al. (1993, 141).

the enzyme, it would take 1.5 metric tons of strong acid 36 hours at high temperatures to digest the same amount of egg whites.

Along with being efficient and extremely reactive, enzymes are characterized by a high degree of specificity. That is, just as a certain key will not fit or unlock each and every lock, enzymes also require an exact molecular fit between the enzyme and the substrate.

Cellular Respiration and Metabolism

The central part of a cell's metabolism is *cellular respiration* (*respiration* is the process by which a compound is oxidized using oxygen as external electron acceptor) because it provides both the raw materials for anabolism and the energy for all other cellular activities. Simply, taken as a whole, *cellular metabolism* is the major chemical reaction that occurs in a cell. It is broken into 2 major divisions, *anabolism* (those reactions that build larger molecules) and *catabolism* (those reactions that break down larger molecules).

METABOLIC TRANSFORMATIONS

Between initial absorption and final excretion, many substances are chemically converted by the organism. The assembly-line activity that occurs in microorganisms during the processing of raw materials into finished products is called **metabolic transformation**. Enzymes mediate metabolic transformations. Note that environmental practitioners must have an understanding of biological processes, including these metabolic transformations. Thus, the following explanation from Spellman and Whiting (2006) is targeted toward environmental practitioners.

Any explanation of a cell or organism's metabolism must include an explanation of the metabolic processes involved. These processes are well known and well documented. As mentioned, the 2 general categories of metabolism are catabolism and anabolism. Again, in catabolic reactions, complex compounds are broken down with a release of energy. These reactions are linked to anabolic reactions, which result in the formation of important molecules. As a result of chemicals and associated reactions, biological cells are dynamic structures that are continually undergoing change.

During metabolism, the cell takes in nutrients, converts them into cell components, and excretes waste into the external environment (see Figure 4.3). Microbial cells are made up of chemical substances, and when the cell grows, these chemical constituents increase in amount. The chemical substances cells need come from the environment; that is, from outside the cell. Once inside the cell, the basic constituents of which the cell is composed transform these substances.

Metabolic reactions require energy for the uptake of various nutrients and for locomotion in motile species. Microorganisms are placed into metabolic classes based on the source of energy they use. In describing these classes, the term *troph* is used (from the Greek meaning "to feed"). Thus, microorganisms that use inorganic

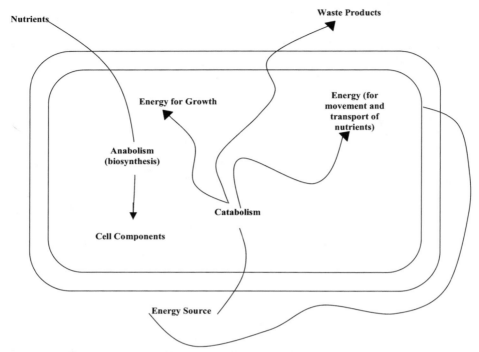

Figure 4.3.   A simplified view of cell metabolism. Adapted from Brock and Madigan (1991).

materials as energy sources are called **lithotrophs** (from the Greek *lithos*, meaning "stone"). Microorganisms that use organic chemicals as energy sources are called **heterotrophs** (feeding from sources other than oneself). Microorganisms that use light as an energy source are called **phototrophs** (from the Greek *photo*, meaning "light"). Most bacteria obtain energy from chemicals taken from the environment and are called **chemotrophs**.

As mentioned, cellular respiration is the process of breaking down food molecules and capturing their energy in ATP molecules. This breakdown process is organized in several steps; all controlled by enzymes. The steps involved include glycolysis, the Krebs cycle and the electron transport system (ETS). Although an in-depth discussion of the metabolic processes of all microorganisms is beyond the scope of this text, environmental practitioners should be well grounded in the fundamental cellular respiration concepts covered briefly in the following sections.

GLYCOLYSIS

Before launching into a discussion of glycolysis, let's pause a moment for a brief review of key material to this point. Recall that several of the chemical reactions that take place in both prokaryotic and eukaryotic cells require an energy input supplied by the molecule ATP. ATP is the high-energy compound whose phosphate groups are easily removed in chemical reactions. ATP supplies energy for 3 main functions:

• To synthesize complex molecules needed by the cell
• To pump substances across the plasma membrane
• For chromosome movement, cilia and flagellar movement and muscle contraction

ATP releases energy by being converted to ADP plus phosphate. ATP is reformed using energy obtained from glycolysis (the subject of the following section) by a coupling reaction between the ADP and a phosphate group.

Glycolysis takes place in the cytoplasm of all cells and is one of the 3 phases of the catabolism of glucose to carbon and water process (the other 2 phases: the **Krebs cycle** and the **electron transport system** are discussed later in this chapter). Glycolysis can occur under both aerobic and anaerobic conditions. Some of the anaerobic processes are called **fermentations**. Fermentation is a process whereby the anaerobic decomposition of organic compounds takes place. These organic compounds serve as both primary and ultimate electron donors and acceptors. Thus, fermentable substances often yield both oxidizable and reductive metabolites (organic compounds produced by metabolism).

The energy-converting metabolism (fermentation) in which the substrate is metabolized without the involvement of an external oxidizing agent is more easily understood by looking at a **metabolic pathway**. For example, in some bacteria, the fermentation of glucose begins with a pathway called glycolysis.

Glycolysis (sometimes referred to as the **Embden-Meyerhof-Parnas pathway—** EMP pathway) involves the breakdown or splitting of glucose (sugar) in a catabolic reaction that concerts 1 molecule of glucose into 2 molecules of the end product, pyruvic acid. In this pathway, energy from energy-yielding (exergonic) reactions is used to phosphorylate ADT—that is, ATP is synthesized from ADP, an example of substrate phosphorylation where energy from a chemical reaction is used directly for the synthesis of ATP from ADP.

The end product in the energy-yielding glycolysis process is the release of a small amount of energy used for various cell functions, and the loss of larger amounts of energy in the form of fermentation products. Common fermentation products of glycolysis include ethanol, lactic acid, alcohols, and gaseous substances that are produced, for example, by certain bacteria.

RESPIRATION

As mentioned, **respiration** is the process by which a compound is oxidized using oxygen as *external* electron acceptor. Using an external electron acceptor is important because in the fermentation process little energy is yielded, mainly because only a partial oxidation of the starting compound occurs in this process. However, if some external terminal acceptor (oxygen, for example) is present, all substrate molecules can be oxidized completely to a by-product (carbon dioxide). When this occurs a far higher yield of ATP is possible.

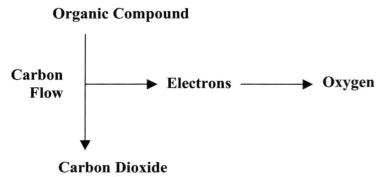

Figure 4.4.   Aerobic respiration—the process by which a compound is oxidized using oxygen as external electron acceptor.

Because an external oxidizing substance is used, the substrate undergoes a net oxidation (see Figure 4.4). The oxidation of a substrate provides more energy than that obtainable from the same substrate with fermentation.

KREBS CYCLE

The **Krebs cycle** (also known as the **tricarboxylic acid cycle**, the **TCA** cycle, or the **citric acid cycle**), commonly called the "energy wheel" of cellular respiration, because it is a cyclical sequence of reactions that takes place in the mitochondrial matrix and is crucial in supplying the energy needs of cells.

When oxygen is available to the cell, the energy in pyruvic acid is released through aerobic respiration. In the TCA cycle, pyruvate is first decarboxylated (the removal of a carboxyl group from a chemical compound), leading to the production of 1 molecule of NADH (NADH is the reduced form of nicotinamide adenine dinucleotide, NAD, which is synthesized in our bodies from the vitamin niacin) and an acetyl coupled to coenzyme A (Acetyl-CoA). The addition of the activated 2-carbon derivative acetyl CoA to the 4-carbon compound oxaloacetic acid forms citric acid, a 6-carbon organic acid. The energy of the high-energy acetyl-CoA bond is used to drive the synthesis. After undergoing dehydration, decarboxylation, and oxidation, 2 additional carbon dioxide molecules are released. Eventually, oxalacetate is regenerated and serves again as an acetyl acceptor, thus completing the cycle. In the course of the cycle, 3 NADHs, 1 FADH (FADH is the reduced form of flavin adenine dinucleotide, FAD—which is synthesized in our bodies from the vitamin riboflavin) and 1 ATP are produced by substrate phosphorylation. The presence of an electron acceptor in respiration allows for the complete oxidation of glucose to carbon dioxide, with a greater yield of energy.

A simplified view or summary of the Krebs cycle is provided as follows:

1. The process begins with the oxidation of pyruvate, producing 1 CO_2, and 1 acetyl-CoA.
2. Acetyl-CoA reacts with the 4-carbon carboxylic acid, oxaloacetate—to form the 6-carbon carboxylic acid, citrate.

3. Through a series of reactions citrate is converted back to oxaloacetate. This cycle produces 2 CO_2 and consumes 3 NAD^+, produce 3 NADH and 3 H^+.
4. It consumes 3 H_2O and consumes 1 FAD, producing 1 $FADH^+$.
5. 1st turn = 1 ATP, 3 NADH, 1 $FADH_2$, 2 CO_2.
6. Since there are 2 molecules of pyruvic acid to deal with, the cycle turns once more.
7. The complete end result = 2 ATP, 6 NADH, 2 $FADH_2$, 4 CO_2.

ELECTRON TRANSPORT SYSTEM (ETS)

The *electron transport system* (ETS) is a common pathway for the use of electrons formed during a variety of metabolic reactions. Most of the electron carriers of the ETS are proteins embedded in the inner mitochondrial membrane (cristae). Most molecules are prevented from going into and out of an organism's cells by the cytoplasmic membrane, the control device of the cellular internal environment. During metabolism, however, the cell must be able to take in various substrates and get rid of waste; this is accomplished by transport systems. In some organisms—Gram-negative bacteria, for example—the transport system is located in membranes other than the cytoplasmic membrane (see Figure 4.5).

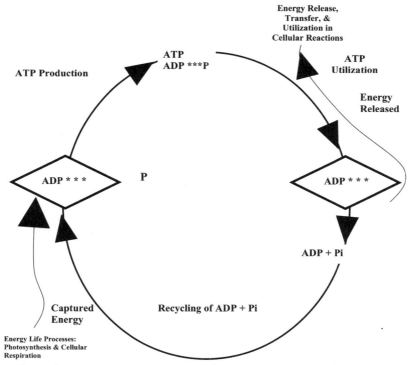

Figure 4.5.   The formation of ATP, a substance that fuels all living organisms by means of phosphorylation, energy-rich bonds are formed, and used to combine ADP and Pi into ATP—which is used to fuel the life processes. Then Pi and ADP are used again in a continuous cycle. Adapted from Wistreich and Lechtman (1980).

A typical ETS is composed of electron carriers. In a bacterium, the ETS involved with respiration occurs in the cytoplasmic membrane. The ETS has 2 functions: to accept electrons from electron donors and transfer them to electron acceptors and to save energy during electron transfer by synthesis of ATP.

The 2 protein components that form the ETS are the **flavoproteins** and **cytochromes**.

Flavoproteins are proteins (enzymes) containing riboflavin, which act as dehydrogenation catalysts or hydrogen-carriers in a number of biological reactions. The flavin portion, which is bound to a protein, is alternately reduced as it accepts hydrogen atoms and oxidized when electrons are passed on. Riboflavin (also called vitamin B2) is a required organic growth factor for some organisms.

Cytochromes are iron-containing proteins that receive and transfer electrons by the alternate reduction and oxidation of the iron atoms, and are important in cell metabolism. Cytochromes in the ETS are known for (among other things) their reduction potentials. One cytochrome can transfer electrons to another with a more positive reduction potential, and can itself accept electrons from cytochromes with a less positive reduction potential.

Photosynthesis

Photosynthesis is the most important biological process on Earth. Basically, in the process of photosynthesis, an organism is able to produce carbohydrates (i.e., sugar, which cellular respiration converts into ATP) and oxygen from water and carbon dioxide. Carbohydrate is synthesized in the process known as the Calvin cycle. The Calvin cycle involves a series of biochemical, enzyme-mediated reactions during which atmospheric carbon dioxide is reduced and incorporated into organic molecules. Eventually some of this forms sugars—oxygen is a by-product of the Calvin cycle. The conversion of unusable sunlight energy into usable chemical energy is associated with the light absorbing actions of the green pigment chlorophyll.

"Photosynthesis is the way plants make fuel molecules to feed their mitochondria. In terms of getting energy, the only real difference between plants and animals is that plants make their own fuel molecules, whereas we animals, in order to get fuel for our mitochondria, must eat something that has eaten plants, or eat plants" (BioMedia 2004).

The basic photosynthesis process uses 6 molecules of water plus 6 molecules of carbon dioxide to produce 1 molecule of sugar plus 6 molecules of oxygen. The overall reaction is written as:

$$6H_2O + 6CO_2 \rightarrow C_6H_{12}O_6 + 6O_2$$

Chapter Review Questions

1. Define the Calvin cycle.
2. Define glycolysis.

3. Define respiration.
4. State the first law of thermodynamics.
5. State the second law of thermodynamics.
6. What does ATP stand for?
7. What does ADP stand for?
8. What does NAD stand for?
9. Explain cofactor.
10. What is a coenzyme?
11. What is the Krebs cycle commonly described as?
12. What is FAD?

References and Additional Reading

Adkins, D.A. *Chapter 12: Meiosis.* Accessed 10/10/06 at http://webpages.marshall.edu/~adkinsda/b120ch12.htm.

Brock, T.D., and M.T. Madigan. 1991. *Biology of Microorganisms.* Englewood Cliffs, NJ: Prentice Hall.

Carroll, L., 2001. *Through the Looking Glass.* New York: Adamant Media Corp.

Eblen, R.A., and W. Eblen. 1994. *The Encyclopedia of the Environment.* Boston: Houghton Mifflin Company.

Farabee, M.J. 2007. *An On-Line Biology Book.* Accessible by www.emc.maricopa.edu/faculty/farabee/biobk/BioBookTOC.html.

Hall, D.O., and K.K. Rao. 1999. *Photosynthesis*, 6th ed. Cambridge, U.K.: Cambridge University Press.

Hoagland, M., and B. Dodson. 1995. *The Way Life Works.* New York: Random House.

Kornberg, A. 1989. *For the Love of Enzymes.* Cambridge, MA: Harvard University Press.

Leninger, D., and M. Cox. 2001. *Principles of Biochemistry*, 3rd ed. New York: Worth Publications.

Lim, D. 1998. *Microbiology*, 2nd ed. New York: William C. Brown/McGraw Hill.

Nave, C.R., 2007. *HyperPhysics: Electrical Work.* Atlanta: Georgia State University.

Prescott, G.W., et al. 1993. *Microbiology*, 3rd ed. Dubuque, IA: Wm. C. Brown Company Publishers.

Russell, B.J. and E.R. Russell. 2004. *How Cells Obtain Energy.* BioMedia Associates. Accessible by http:ebiomedia.com/prod/LC/LCenergy.html.

Spellman, F.R., and N. Whiting. 2006. *Environmental Science and Technology*, 2nd ed. Lanham, MD: Government Institutes.

Thomas, L. 1978. *Lives of a Cell: Notes of a Biology Watcher.* New York: Penguin.

Wistreich, G.A., and M.D. Lechtman. 1980. *Microbiology.* 3rd ed. New York: Macmillan.

Witkowski, A., and J. Power. 1975. *Enzymes: Natures Catalysts.* Burlington, NC: Carolina Biological Co.

CHAPTER 5

The Cell Cycle and Genetics

Statistically the probability of any one of us being here is so small
that you would think the mere possibility of existence would keep
us all in a contented dazzlement of surprise.

—Lewis Thomas (1913–1993)

Topics in This Chapter

- The Cell Cycle
- Making New Cells
- Genetics
- DNA

In the last quarter century there has been a revolution in our comprehension of how cells grow and divide. Results from experiments on embryos, yeast, and cultured mammalian cells have unified disparate viewpoints into a single set of principles of normal cellular reproduction in plants, animals, and bacteria (Murray & Hunt 1993).

Approximately 25 million cell divisions occur in an adult human every second. The cells of our hair follicles, skin, bone marrow, and the lining of our GI tracts are turning over, dividing constantly. Why? That is, why must cells divide? Isn't growth a characteristic of all living things? And since this is the case, why can't cells just continue to grow along with everything else? Cells must divide because total size of a cell is limited. This is because, as cell size increases, the surface area to volume ratio decreases (i.e., it becomes harder for materials to diffuse to the center of the cell). Moreover, larger cells would need more DNA to make proteins than one nucleus can contain. Accordingly, multicellular organisms grow by cell division. Cell growth and division are usually controlled by the organism for reproduction (in bacteria and amoeba) to allow an organism to grow and develop, and to repair or replace damaged or dead cells.

✓   **Important Point**: Uncontrolled cell division = cancer.

The Cell Cycle

The **cell cycle** is an ordered set of events. These events or phases, diagrammed below, culminate in cell growth and division into 2 daughter cells—nondividing cells are not considered to be in the cell cycle.

INTERPHASE

Period of *cell growth* between cell divisions. Includes phases 1–3:

Phase 1: G1 (Gap 1)
Cell growth and development

Phase 2: S (Synthesis)
DNA *replication* takes place (result is 2 copies of each DNA molecule at end of phase)

Phase 3: G2 (Gap 2)
Organelles and other materials necessary for cell division are synthesized

MITOSIS

Phase 4: M (Mitosis)
Period of *nuclear division*

Making New Cells

According to the National Center for Biotechnology Information (2006), for most unicellular organisms, reproduction is a simple matter of **cell duplication**, also known as **replication**. But for multicellular organisms, cell replication and reproduction are 2 separate processes. Multicellular organisms replace damaged or worn out cells through a replication process called **mitosis**, the division of a eukaryotic cell nucleus to produce 2 identical **daughter nuclei**. To reproduce, eukaryotes must first create special cells called **gametes**—eggs and sperm—that then fuse to form the beginning of a new organism. Gametes are but one of the many unique cell types that multicellular organisms need to function as a complete organism.

BINARY FISSION

Most unicellular organisms create their next generation by replicating all of their parts and then splitting into 2 cells, a type of **asexual reproduction** called **binary fission**. This process spawns not just 2 new cells, but also 2 new organisms. Multicellular organisms replicate new cells in much the same way. For example, we produce new skin cells and liver cells by replicating the DNA found in that cell through mitosis. Yet, producing a whole new organism requires **sexual reproduction**, at least for most multicellular organisms. In the first step, specialized cells called **gametes**—eggs and sperm—are created through a process called **meiosis**. Meiosis serves to reduce the chromosome number for that particular organism by half. In the second step, the sperm and egg join to make a single cell, which restores the chromosome number. This joined cell then divides and differentiates into different cell types that eventually form an entire functioning organism (NIH 2004).

KEY MITOSIS/MEIOSIS TERMS

The process of mitosis and meiosis are diagrammed below. Before studying the mitosis/meiosis breakdowns, review the definitions of key terms:

- **Centriole**—a cylindrical cytoplasmic organelle located just outside the nucleus of animal cells and the cells of some lower plants; associated with the spindle during mitosis and meiosis.
- **Centromere**—a special region on a chromosome from which kinetochore microtubules radiate during mitosis or meiosis.
- **Chromatin**—the mixture of DNA and protein (mostly histones in the form of nucleosome cores) that comprises eukaryotic nuclear chromosomes.
- **Chromosomes**—a filamentous structure in the cell nucleus, mitochondria, and chloroplasts along which the genes are located. The chromosome characteristics of humans include:
  - A diploid set (in which cells contain 2 sets of chromosomes,1 maternal and 1 paternal); abbreviated 2n; 2n = 46
  - Autosomes (chromosomes that are not sex chromosomes); these are homologous chromosomes, 1 from each parent (22 sets of 2)
  - Sex hormones (humans have 1 set of 2):
    1. Female—sex chromosomes are homologous (XX)
    2. Male—sex hormones are nonhomologous (XY)
- **Histones**—one of a class of basic proteins serving as structural elements of eukaryotic chromosomes.
- **Kinetochore**—placed on either side of the centromere to which the spindle fibers are attached during cell division.
- **Nucleosome**—a complex consisting of several histone proteins, which together form a "spool," and chromosomal DNA, which is wrapped around the spool.
- **Spindle**—a microtubular structure with which the chromosomes are associated in mitosis and meiosis.
- **Tetrad**—a homologous pair of double-stranded chromosomes, attached at the centromeres.

STAGES OF MITOSIS

Mitosis is the period of nuclear division. Every time a cell divides, it must ensure that its DNA is shared between the 2 daughter cells. Mitosis is the process of "divvying up" the genome between the daughter cells:

- **Interphase**

    - Technically not a part of mitosis, but it is included in the cell cycle
    - Cell is in a resting phase, performing cell functions
    - DNA replicates (copies)
    - Organelles double in number, to prepare for division

- **Prophase**

    - Chromatin coils into short thick structures (known as *chromosomes*) and become visible
    - *Centrioles* separate and move to opposite sides of the nucleus (animal cells only)
    - Paired chromosomes (*chromatids*) attach to *spindle fibers* at *centromere* (spot where chromatids are attached to each other)
    - Nuclear membrane and nucleolus break down and disappear

- **Metaphase**

    - Chromosomes line up across the *equator*, or center of the cell, pulled by the spindle fibers

- **Anaphase**

    - Centromeres split, sister chromatids separate, are drawn to opposite *poles* of the cell by the spindle fibers

- **Telophase**

    - Chromosomes begin to uncoil
    - Nuclear envelopes form around 2 daughter nuclei
    - Nucleolus forms in each nucleus
    - Spindle fibers break apart
    - Cytoplasm begins to divide (*cytokinesis*)

- **Cytokinesis (final step)**

    - Cell's cytoplasm divides in 2 ways:

        Animal cells—cell membrane moves in towards the center of cell, cell pinched into 2 nearly equal parts
        Plant cells—cell plate forms midway between the 2 nuclei
    - Cell wall forms on either side of cell plate

✓   **Important Point**: Cytokinesis happens differently in plant cells because it isn't easy to pinch off that tough cell wall.

MEIOSIS

Meiosis is a specialized type of cell division that occurs during the formation of gametes. That is, meiosis involves reproduction. Although meiosis may seem much more complicated than mitosis, it is really just 2 cell divisions in sequence. Each of these sequences maintains strong similarities to mitosis. Meiosis differs in that this single replication is followed by 2 consecutive divisions—Meiosis I and Meiosis II; meiosis produces 4 daughter cells and each has half the number of chromosomes as the original cell. Dean A. Adkins of the website "Biology Notes" describes the stages of meiosis:

Stages of Meiosis I

- **Interphase I**

  - Chromosomes replicate as in mitosis

- **Meiosis I**

  - Reduces chromosome number by one half

- **Prophase I**

  - Chromosomes condense
  - Synapse occurs; during this process homologous chromosomes come together as pairs. Each homologous pair of replicated chromosome of 2 chromatids complex as 4 intertwined chromatids—*tetrad*. Nonsister chromatids are linked by X-shaped chiasmata, sites where homologous strand exchange or crossing-over occurs
  - Other event similar to prophase of mitosis

- **Metaphase I**

  - Tetrads are aligned on the metaphase plate
  - Each homologue attached to kinetochore microtubules

- **Anaphase I**

  - Chromosomes are moved toward poles by spindle apparatus
  - Sister chromatids remain attached at their centromeres
  - Differs from mitosis in that chromatids do not separate at this time

- **Telophase I and Cytokinesis**

  - Chromosomes reach poles
  - Each pole now has a haploid set of chromosomes still composed of 2 sister chromatids attached at the centromere
  - Cytokinesis occurs; some cells pause but others immediately prepare for Meiosis II

Stages of Meiosis II

- **Prophase II**

  - Nuclear envelope and nucleoli disperse
  - Spindle apparatus forms

- **Metaphase II**

  - Chromosomes align singly on metaphase plate

- **Anaphase II**

  - Centromeres of sister chromatids separate
  - Sister chromatids of each pair (now individual chromosomes) move toward opposite poles of the cell

- **Telophase and Cytokinesis**

  - Nuclei form
  - Cytokinesis occurs, producing 4 haploid daughter cells (Adkins 2006)

KEY DIFFERENCES BETWEEN MITOSIS AND MEIOSIS

Meiosis is reduction division. Cells produced by mitosis have same number of chromosomes as the original cell, but cells produced by meiosis have half the number of chromosomes as the original cell. Meiosis creates genetic diversity. Mitosis produces 2 cells identical to parent cell whereas meiosis produces 4 cells genetically different from parent cell and from each other. Meiosis is 2 successive nuclear divisions.

Genetics

When Gregor Mendel, an Austrian monk, began his hybridization experiments with pea plants in 1856, knowledge of how heredity works was limited. For example, if black furred animals mated with white furred animals, it was expected that all resulting progeny would be gray (a color intermediate between black and white). They found, of course, that this is often not the case. The correct explanation of inherited traits was provided by Mendel (and later by others) through his study of peas.

KEY GENETIC TERMS

Before discussing Mendelian genetics it is important to understand the context of his times as well as how his work fits into the modern science of genetics. Key genetic terms include:

- **Allele**—an alternate form of a gene. Usually there are 2 alleles for every gene, sometimes as many as 3 or 4.
- **Autosome**—a chromosome that is not a sex chromosome.
- **Chromosome**—a structure in the cell nucleus that carries the genes and is capable of reproduction through cell division.
- **Dihybrid cross**—tracking the inheritance of 2 traits between 2 individuals.
- **Dominant trait**—term applied to the trait (allele) that is expressed irregardless of the second allele.
- **Gene**—unit of inheritance that usually is directly responsible for one trait or character.
- **Genotype**—the allelic composition of an organism.
- **Heterozygous**—when the 2 alleles are different, in such cases the dominant allele is expressed.
- **Homozygous**—when the 2 alleles are the same.
- **Monohybrid cross**—tracking the inheritance pattern of a single trait between 2 individuals.
- **Phenotype**—expressed traits of an individual.
- **Punnett squares**—probability diagram illustrating the possible offspring of a mating.
- **Recessive trait**—term applied to a trait that is only expressed when the second allele is the same.
- **Test cross**—a mating between an individual with an unknown genotype for a specific trait with an individual who is homozygous recessive for the same trait.
- **True-breeding**—sexually reproduced organisms with inherited traits(s) identical to parents.

✓   **Important Point**: A *gene* is a discrete unit of heredity located on chromosomes that consists of DNA. An *allele* is an alternate form of a specific gene. (e.g., eye color gene vs. blue eye allele)

MENDELIAN GENETICS

In 1856, Mendel developed the fundamental principles that would become the modern science of genetics. Mendel was successful in his studies because he used a quantitative approach; counted offspring traits over several generations; used probabilities to interpret the results; studied one trait at a time; and used the common garden pea (*Pisum sativum*). The common garden pea is easy to cultivate and has a short generation time. Mendel examined 7 traits in the pea plant: flower color, flower position, seed color, seed shape, pod color, pod shape, and stem length. Mendel reasoned that each characteristic only occurs in 2 contrasting forms—dominant and recessive.

Mendel chose the common garden pea because he knew that pollen from a pea plant can fertilize the female eggs of the same plant. Self-fertilization provides strict control over mating. That is, seeds produced by self-fertilization inherit only traits present in the plant that bore them. With continued experimentation, Mendel learned how to prevent self-fertilization and performed many cross-pollination experiments.

Monohybrid Crosses

A hybrid is the product of parents that are true-breeding for distinctly different traits. A **monohybrid cross** is a cross involving a single trait. Mendel tracked traits through 2 generations:

P generation—parental
F1 generation—first offspring
F2 generation—second generation obtained from cross-breeding F1 generation

The results of Mendel's experiments contradicted those of previous theories. He found that F1 resembled parents and F2 showed 1/4 plants resembled one parent, 3/4 resembled the other.

✓   **Important Point**: Mendel found that phenotypes (genetically determined appearance) were in a 3:1 ratio (Grobner 2004).

Dihybrid Crosses

Mendel used **dihybrid crosses** to explain how 2 pairs of genes are distributed in the gametes. Mendel crossed tall peas that had green pods and short peas that had yellow pods. All the offspring of the F1 generation had the same phenotype—tall pea plants with green pods.

When he allowed the hybrid pea plant to self-fertilize, he found a phenotypic relationship of 9:3:3:1 among the plants in the F2 generation (Grobner 2004).

Theory of Segregation

During his era, Mendel did not know about genes and meiosis. However, his Theory of Segregation still stands. During several experiments, Mendel developed a 4-part hypothesis: (1) he suggested that alternative forms of inherited units are responsible for variations in traits; (2) for each trait, an individual inherits 2 alleles, one from each parent; (3) if the alleles differ, one is expressed (dominant) and the other is masked (recessive); and (4) the 2 alleles segregate during gamete production.

Theory of Independent Assortment

As a result of his dihybrid crosses, Mendel thought that inheritance of a pair of factors for one trait is independent of the simultaneous inheritance for another trait. Today, we know that this theory can be explained by meiosis—factors assort independently.

Types of Inheritance

1. **Incomplete dominance**

   - Offspring show traits intermediate between 2 parental phenotypes
   - Red-flowered and white-flowered four o'clocks produce pink flowers
   - Not a blending—parental phenotypes reappear in F2

2. **Codominance**

   - Both alleles of a gene are expressed
   - Both genes produce an effective allele
   - Human blood types—AB individuals' red blood cells (RBCs) express both A and B antigens

3. **Multiple alleles**

   - More than 2 alleles for a given locus
   - Only 2 alleles inherited
   - ABO blood system

     - Alleles A, B, and O
     - Blood types A, B, AB, and O

4. **Epistasis** (masking of the effects of one gene by the action of another)

   - Absence of expected phenotype due to masking on one gene pair by another
   - Homozygous recessive at one locus masks the effect of a dominant allele at another
   - Albino animals—inherit allelic pair (aa) preventing melanin production
   - Mammal coat color

5. **Pleiotropy** (a genotype with multiple phenotypic effects)

   - Single gene exerts on many aspects of an individual's phenotype

6. **Polygenic inheritance**

   - A trait is controlled by several alleic pairs at different loci
   - Gene alleles can be contributing or noncontributing
   - Contributing alleles have additive effect resulting in variation
   - Subject to environmental effects producing intermediate phenotypes

7. **Environmental effects on the phenotype** (the outward appearance of an organism)

   - Both genotype and environment affect phenotype
   - Relative importance of each varies
   - Calculated as heritabilities
   - Intelligence has environmental component

POST-MENDEL GENETICS

Mendel published his research in 1865. However, his unique, groundbreaking notion of genes was not appreciated by naturalists of his time. Thus, Mendel's work lay fallow until 1900, when scientists independently confirmed his results. These confirmations, based on the study of the cell and chromosomal behavior, gave Mendel's abstract work the physical context—the teeth—it needed. For example, modern genetics explain Mendel's observations in the following ways:

- **Each trait is controlled by 2 alleles**

  - Alternative gene forms at the same gene locus on homologous chromosomes:

    Dominant allele—masks expressions of recessive allele
    Recessive allele—only expressed when both alleles are recessive

- **Gene locus**

  - Specific location of a particular gene on homologous chromosomes
  - Mendel's true-breeding plants were homozygous for the traits he studied:

    Homozygous—contains the same alleles at both loci
    Heterozygous—contain different alleles at the 2 loci

  - Homozygous dominant:

    Possess 2 dominant alleles for a trait

  - Homozygous recessive

    Possess 2 recessive alleles for a trait
    Trait is expressed

  - In heterozygous organisms, the recessive trait is not expressed

- **Genotype vs. phenotype**

  - Genotype:

    Refers to the genetic makeup of an individual
    Lists the alleles present

  - Phenotype:

    Refers to the physical appearance of the individual
    Represents gene expression

- **Punnett square**

  - Provides a simple method to determine the probable offspring of a genetic cross
  - Used to show all the possible outcomes of a genetic cross and to determine the probability of a particular outcome.

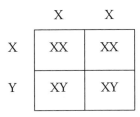

- In the Punnett square shown above, the XX chromosomes represent a female and XY represents a male. From the Punnett square it can be seen that scientists can figure out the percentage of an offspring being a boy or girl with specific traits.

- **Test crosses**
  - Allow for the verification of genotypes in an individual
  - Dominant individual may be heterozygous or homozygous dominant
  - Use a homozygous recessive individual as the test
  - Recessive phenotypes (½) in the offspring indicate heterozygous individual

HUMAN GENETICS

There are 44 *autosomes* and 2 sex *chromosomes* in the human male and female genome, for a total of 46. There may be only 2 sex chromosomes, but they are critically important. They determine the gender of the individual. Females have XX chromosomes and males have XY chromosomes. The Y chromosome is much smaller than the X.

DNA

NIH (2006) points out that DNA, or deoxyribonucleic acid, is the hereditary material in humans and almost all other organisms. Nearly every cell in a person's body has the same DNA. Most DNA is located in the cell nucleus (where it is called nuclear DNA), but a small amount of DNA can also be found in the mitochondria (where it is called mitochondrial DNA or mtDNA).

The information in DNA is stored as a code made up of 4 chemical bases: adenine (A), guanine (G), cytosine (C), and thymine (T). Human DNA consists of about 3 billion bases, and more than 99% of those bases are the same in all people. The order, or sequence, of these bases determines the information available for building and maintaining an organism, similar to the way in which letters of the alphabet appear in a certain order to form words and sentences.

DNA bases pair up with each other—A with T and C with G—to form units called base pairs. Each base is also attached to a sugar molecule and a phosphate molecule. Together, a base, sugar, and a phosphate are called a nucleotide. Nucleotides are arranged in 2 long strands that form a spiral called a double helix. The structure of the double helix is somewhat like a ladder, with the base pairs forming the ladder's rungs and the sugar and phosphate molecules forming the vertical sidepieces of the ladder.

An important property of DNA is that is can replicate, or make copies of itself. Each strand of DNA in the double helix can serve as a pattern for duplicating the sequence of bases. This is critical when cells divide because each new cell needs to have an exact copy of the DNA present in the old cell (NIH 2006).

Chapter Review Questions

1. What kind(s) of cell division do eukaryotes do?
2. To which part of a chromosome do spindle fibers attach?
3. What are the stages of the cell cycle?
4. What are the stages of mitosis?
5. How many cells are present at the end of mitosis?
6. What are the stages and phases of meiosis?
7. What types of chromosomes form a tetrad?
8. How many cells are present at the end of meiosis?
9. What makes meiosis different from mitosis?
10. Define and distinguish between the terms "gene" and allele."
11. Define and distinguish between the terms "genotype" and "phenotype."
12. Define the terms "dominant" and "recessive."

References and Additional Reading

Grobner, M.A. 2004. *Biology 1010: Genetics Lecture.* Stanislaus, CA: California State University.

John, B., et al. 1990. *Meiosis.* London: Cambridge University Press.

Mitosis/Meiosis DVD. 2004. New York: Educational Video Network, Inc.

Murray, A.W., and T. Hunt, 1993. *Cell Cycle: An Introduction.* New York: W.H. Freeman.

NCBI. 2006. *A Science Primer.* National Center for Biotechnology Information. Accessed July 6, 2006, at www.ncbi.nim.nih.gov.

NIH. 2004. *A Science Primer: Making New Cells.* Washington, DC: Department of Health and Human Services, National Institutes of Health.

NIH. 2006. What Is DNA. Washington, DC: National Institutes of Health. Accessed August 19, 2006, at http://ghr.nlm.nih.gov/handbook/basics/dna.

Parker, G., and W.A. Reynolds. 1979. *Mitosis and Meiosis.* Illinois: Dearborn Trade Publishing.

Rieder, C.L., et al. 1998. *Mitosis and Meiosis: Methods in Cell Biology,* Vol. 6. New York: Academic Press.

Thomas, L. 1995. *The Medusa and the Snail: More Notes of a Biology Watcher.* New York: Penguin Books.

CHAPTER 6

Biological Diversity

The most useful piece of learning for the uses of life is to unlearn what is untrue.

—Antisthenes

Topics in This Chapter

- Classification
- Kingdoms of Life
- Bacteria
- Viruses
- Protists
- Fungi
- Plants
- Animals

  - Invertebrates
  - Vertebrates

- Human Evolution

Biological diversity, or "biodiversity," refers to the variety of genes, species, and ecosystems found on Earth.

The National Institutes of Health (NIH) (1994) makes the point that perhaps the best way to illustrate the importance of biodiversity is by analogy to the diversity of human knowledge stored in books (an argument made by Tom Lovejoy and others). When the library in Alexandria was destroyed in AD 391, when Constantinople was sacked in 1453, or when Maya codices were burned in the 16th century, thousands of works of literature were destroyed. Hundreds of works of genius are now known to us only by their titles, or from quoted fragments. Thousands more will never be known; several millennia of collective human memory have been irretrievably lost.

Like books, living species represent a kind of memory, the cumulative record of several *million* millennia of evolution. Every species has encountered and survived countless biological problems in its evolutionary history; molecules, cells, and tissues record their solutions. Because we are biological beings ourselves, nature offers a vast library of solutions to many of our current health, environmental, and economic problems. Unfortunately, that precious and irretrievable information is now being destroyed at an unprecedented rate (NIH 1994).

Fast-forward to the present. Population pressures and demographic changes threaten biodiversity worldwide. Our cities produce a growing stream of industrial poisons and human wastes. In the countryside, overproduction depletes the soil, and pesticides contaminate water supplies; deforestation for farming, pasture, and building material leads to erosion and heavy flooding. The resulting contraction of natural habitats—including the destruction of species-rich tropical forests—will have profound consequences for the future (NIH 1994).

We have identified more than 2 million species of organisms on Earth, but estimate 40 million species inhabit it. Some estimate that there may be millions of species in the tropical rain forest and an unspecified number living in the oceans yet undiscovered. Then there is this: Of the millions of species presently identified, how many do we really understand, know, or can we accurately explain? When studying the various species, it is important to remember the words of Ralph Waldo Emerson (1889): "What is a weed? A plant whose virtues have not yet been discovered."

Classification

For centuries, scientists classified the forms of life visible to the naked eye as either animal or plant. The Swedish naturalist Carolus Linnaeus organized much of the current knowledge about living things in 1735.

The importance of classifying organisms cannot be overstated. Without a classification scheme, how could we establish criteria for identifying organisms and arranging similar organisms into groups? The most important reason for classification is that a standardized system allows us to handle information efficiently—it makes the vastly diverse and abundant natural world less confusing.

Linnaeus's classification system was extraordinarily innovative. His **binomial system of nomenclature** is still with us today. Under the binomial system, all organisms are generally described by a 2-word scientific name, the **genus** and **species**. Genus and species are groups that are part of a hierarchy of groups of increasing size, based on their nomenclature (taxonomy). This hierarchy is:

Kingdom
　　Phylum
　　　　Class
　　　　　　Order
　　　　　　　　Family
　　　　　　　　　　Genus
　　　　　　　　　　　　Species

Using this hierarchy and Linnaeus's binomial system of nomenclature, the scientific name of any organism (as stated previously) includes both the genus and the species name. The genus name is always capitalized, while the species name begins with a lowercase letter. On occasion, when little chance of confusion is present, the genus name is abbreviated to a single capital letter. The names are always in Latin, so they are usually printed in italics or underlined. Some organisms also have English common names. Some microbe names of interest, for example, are listed as follows:

- *Salmonella typhi*—the typhoid bacillus
- *Escherichia coli*—a coliform bacterium
- *Giardia lamblia*—a protozoan

✓   **Interesting Point**: *Escherichia coli* is commonly known as simply *E. coli*, while *Giardia lamblia* is usually referred to by only its genus name, *Giardia*.

Kingdoms of Life

Linnaeus classified all then-known (1700s) organisms into 2 large groups: the kingdoms Plantae and Animalia. In 1969, Robert Whittaker proposed 5 kingdoms: Monera, Protista, Fungi, Plantae, and Animalia. Other schemes involving an even greater number of kingdoms have lately been proposed, however this text employs Whittaker's 5 kingdoms. Moreover, recent studies suggest that 3 domains (super-kingdoms) be employed: Archaea, Bacteria, and Eukarya.

The basic characteristics of each kingdom are summarized in the following:

1. **Kingdom Monera** (10,000 species)—unicellular and colonial—including **archaebacteria** (from the Greek *archae* meaning "ancient") and **eubacteria** (from the Greek *eu* meaning "true"). Archaebacteria include methanogens (producers of methane), halophiles (live in bodies of concentrated salt water) and thermocidophiles (live in the hot acidic waters of sulfur springs). Eubacteria include heterotrophs (decomposers), autotrophs (make food from photosynthesis), and proteobacteria (one of the largest phyla of bacteria). All prokaryotic cells (without nuclei and membrane-bound organelles) are in this kingdom. Both reproduce by binary fission, but they do have some ways to recombine genes, allowing change (evolution) to occur.
2. **Kingdom Protista** (250,000 species)—unicellular protozoans and unicellular and multicellular (macroscopic) algae with cilia and flagella. Kingdom Protista contains all eukaryotes that are not plants, animal or fungi. Includes *Amoebae* and *Euglena*.
3. **Kingdom Fungi** (100,000 species)—multicellular and heterotrophic eukaryotes having multinucleated cells enclosed in cells with cell walls. Fungi act either as decomposers or as parasites in nature. Includes molds, mildews, mushrooms, and yeast.
4. **Kingdom Plantae** (250,000 species)—immobile, multicellular eukaryotes; carry out photosynthesis (autotrophs) and have cells encased in cellulose cell walls. Plants are important sources oxygen, food, and clothing/construction materials, as well as pigments, spices, drugs and dyes.

5. **Kingdom Animalia** (1,000,000 species)—multicellular, heterotrophic eukaryotes; without photosynthetic pigment, mostly move from place to place. Animal cells have no cell walls.

✓   **Important Point**: Note that recent practice is to place archaebacteria in a separate kingdom, Kingdom Archaebacteria. This is the case because data from DNA and RNA comparisons indicate that archaebacteria are so different that they should not even be classified with bacteria. Thus, a separate and distinct classification scheme higher than kingdom has been devised to accommodate the archaebacteria, called **domain**. In this new system, these organisms are now placed in the domain **Archaea**—the chemosynthetic bacteria. Other prokaryotes, including eubacteria, are placed in the domain **Bacteria**—the disease-causing bacteria. All the kingdoms of eukaryotes, including Protista, Fungi, Plantae and Animalia, are placed in the domain **Eukarya**.

✓   **Note**: A brief overview of selected organisms from each of the 5 major kingdoms is presented in the following sections.

Bacteria

Of all organisms, bacteria are the most widely distributed, the smallest in size, the simplest in morphology (i.e., structure; see Figure 6.1), the most difficult to classify, and the hardest to identify. Because of their considerable diversity, even providing a descriptive definition of what a bacterial organism is can be difficult. About the only generalization that can be made for the entire group is that they are single-celled, prokaryotic, seldom photosynthetic, and reproduce by binary fission.

Bacteria cells are usually measured in microns, μ, or micrometers, μm; 1 μm = 0.001 or $\frac{1}{1000}$ of a millimeter, mm. A typical coliform bacterial cell that is rod-shaped is about 2 μm long and about 0.7 microns wide. The size of each cell changes with time during growth and death.

The arrangement of bacterial cells, viewed under the microscope, may be seen as separate (individual) cells or as cells in groupings. Within their species, cells may appear in pairs (diplo), chains, groups of four (tetrads), cubes (Sarcinae), and in clumps. Long chains of cocci result when cells adhere after repeated divisions in 1 plane; this pattern is seen in the genera *Enterococcus* and *Lactococcus*. In the genus *Sarcina*, cocci divide in 3 planes, producing cubical packets of 8 cells. The shape of rod-shaped cells varies, especially the rod's end, which may be flat, cigar-shaped, rounded, or bifurcated. While many rods do occur singly, they may remain together after division to form pairs or chains (see Figure 6.2). These characteristic arrangements are frequently useful in bacterial identification.

Bacteria are found everywhere in our environment. They are present in soil, in water, and in the air. Bacteria are also present in and on the bodies of all living creatures—including people. Most bacteria do not cause disease; they are not pathogenic. Many bacteria carry out useful and necessary functions related to the life of larger organisms.

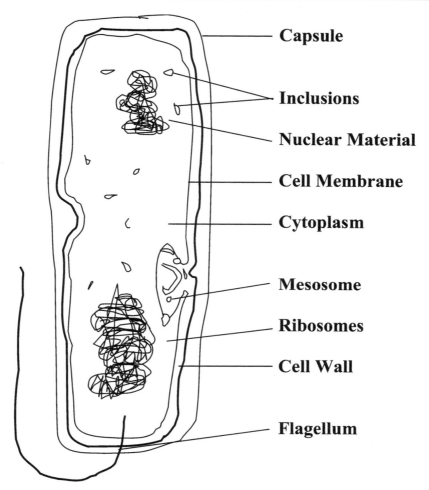

Figure 6.1. Bacterial cell.

However, when we think about bacteria in general terms, we usually think of the damage they cause. In water, for example, the form of water pollution that poses the most direct menace to human health is bacteriological contamination, part of the reason that bacteria are of great significance to water and wastewater specialists. For water treatment personnel tasked with providing the public with safe, potable water, disease-causing bacteria pose a constant challenge (see Table 6.1).

The conquest of disease has placed bacteria high on the list of microorganisms of great interest to the scientific community. There is more to this interest and accompanying large research effort than just an incessant search for understanding and the eventual conquest of disease-causing bacteria. Not all bacteria are harmful to man. Some, for example, produce substances (antibiotics) that help in the fight against disease. Others are used to control insects that attack crops. Bacteria also have an impact on the natural cycle of matter. Bacteria work to increase soil fertility, which increases

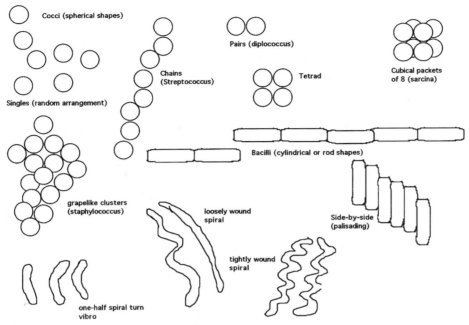

Figure 6.2.   Bacterial shapes and arrangements.

the potential for more food production. With the burgeoning world population, increasing future food productivity is no small matter.

We have a lot to learn about bacteria, because we are still principally engaged in making observations and collecting facts, trying wherever possible to relate one set of facts to another, but lacking much of a basis for grand unifying theories. Like most learning processes, gaining knowledge about bacteria is a slow and deliberate process. With more knowledge about bacteria, we can minimize their harmful potential and exploit their useful activities.

Bacteria come in 3 shapes: elongated rods called **bacilli**, rounded or spherical cells called **cocci**, and spirals (helical and curved) called **spirilla** (for the less rigid form) and **spirochaete** (for those which are flexible). Elongated rod-shaped bacteria may vary considerably in length; have square, round, or pointed ends; and may be motile (possess the ability to move) or nonmotile. The spherical-shaped bacteria may occur singly,

Table 6.1.   Disease-Causing Bacterial Organisms Found in Polluted Water

Microorganism	Disease
Salmonella Typhi	typhoid fever
Salmonella sp.	salmonellosis
Shigella sp.	shigellosis
Campylobacter jejuni	campylobacter enteritis
Yersinia entercolitice	yersiniosis
Escherichia coli	E. coli

Elaborate Irregular

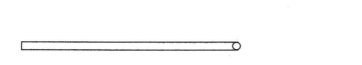

Long Slender Rod   **Geometric Polyhedral**

Figure 6.3.   Virus shapes.

in pairs, in tetrads, in chains, and in irregular masses. The helical and curved spiral-shaped bacteria exist as slender spirochaetes, spirillum, and bent rods (see Figure 6.3).

Viruses

Viruses are parasitic intracellular particles that are the smallest living infectious agents known. They are not cellular—they have no nucleus, cell membrane, or cell wall. Viruses, like cells, carry genetic information encoded in their nucleic acid, and can undergo mutations and reproduce; however, they cannot carry out metabolism, and thus are not considered alive. They multiply only within living cells (hosts) and are totally inert outside of living cells, but can survive in the environment. Just a single virus cell can infect a host. As far as measurable size goes, viruses range from 20 to 200 millimicrons in diameter, about 1 to 2 orders of magnitude smaller than bacteria. More than 100 virus types excreted from humans through the enteric tract could find their way into sources of drinking water. In sewage, these average between 100 to 500 enteric infectious units/1000 ml. If the viruses are not killed in various treatment processes and become diluted by a receiving stream, for example, to 0.1–1 viral infectious units/100 ml, the low concentrations make it very difficult to determine virus levels in water supplies. Since tests are usually run on samples of less than 1 ml, at least 1,000 samples would have to be analyzed to detect a single virus unit in a liter of water.

Viruses differ from living cells in at least 3 ways: (1) they are unable to reproduce independently of host cells and carry out cell division; (2) they possess only one type of nucleic acid, either DNA or RNA; and (3) they have a simple cellular organization. Some viruses that may be transmitted by water include hepatitis A, adeno virus (a DNA virus that causes colds and "pink eye"), polio, Coxsackie, echo and Norwalk agent. A virus that infects a bacterium is called a **bacteriophage**.

Lewis Thomas, in *The Lives of a Cell*, points out that when humans "catch diphtheria it is a virus infection, but not of us." That is, when humans are infected by the virus causing diphtheria, it is the bacterium that it really infects—humans simply "blundered into someone else's accident" (1974, 76). The toxin of diphtheria bacilli is produced when the organism has been infected by a bacteriophage.

✓   **Interesting Point**: The Papillomavirus is a DNA virus that causes warts. These infectious particles are small, about 15 nm in diameter.

A bacteriophage (phage) is any viral organism whose host is a bacterium; that is, any organisms that infects bacteria. Most of the bacteriophage research that has been carried out has been on the bacterial *Escherichia coli*, one of the gram-negative bacteria that environmental specialists such as water and wastewater operators are concerned about because it is a dangerous typical coliform (Spellman 2000).

A virus does not have a cell-type structure from which it is able to metabolize or reproduce. However, when the **genome** (a complete haploid set of chromosomes) of a virus is able to enter into a viable living cell (a bacterium) it may "take charge" and direct the operation of the cell's internal processes. When this occurs, the genome, through the host's synthesizing process, is able to reproduce copies of itself, move on, and then infect other hosts. Hosts of a phage may involve a single bacterial species or several bacteria genera.

The most important properties used in classifying bacteriophages are nucleic acid properties and phage morphology. That is, viruses are classified by the type of nucleic acid they contain, and the shape of their protein capsule (**capsid**). Bacterial viruses may contain either DNA or RNA; most phages have double-stranded DNA (Spellman 2000).

✓   **Interesting Point**: The Influenza virus causes the flu. It has RNA as its genetic material instead of DNA.

Many different basic structures have been recognized among phages. Phages appear to show greater variation in form than any other viral group. (Basic morphological structures of various viruses are shown in Figure 6.3.) The T-2 phage virus has 2 prominent structural characteristics: the head (a polyhedral capsid) and the tail.

The effect of phage infection depends on the phage and host and to a lesser extent on conditions. Some phages multiply with and **lyse** (destroy) their hosts. When the host lyses (dies and breaks open), phage progeny are released.

✓ **Important Point**: Bacteriophages invade the host cell, take over the cell, and begin replicating viruses, eventually lysing or bursting the host cell, releasing the new viruses to infect additional cells (Spellman 2000).

"Viruses cause a variety of diseases among all groups of living organisms. Viral diseases include the flu, common cold, herpes, measles, chicken pox, small pox, and encephalitis. Antibiotics are not effective against viruses. Vaccination offers protection for uninfected individuals" (Farabee 2006).

Protists

Protists are unicellular and multicellular eukaryotes, which exhibit a great deal of variation in their life cycles. The Protists include heterotrophs, autotrophs, and some organisms that can vary their nutritional mode depending on environmental conditions. Protists occur in freshwater, saltwater, soil, and as symbionts within other organisms; they include protozoa, algae, and slime molds. Because of this tremendous diversity, classification of the Protista is difficult.

✓ **Important Point**: Protists are not plants, animals, or fungi, but they act enough like them that scientist believe protists paved the way for the evolution of early plants, animals, and fungi.

PROTOZOA

The **protozoa** ("first animals" or "little animals") are a large group of eukaryotic organisms (more than 50,000 known species that have adapted a form or cell to serve as the entire body). All protozoans are single-celled organisms. Typically they lack cell walls, but have a plasma membrane that is used to take in food and discharge waste. They can exist as solitary or independent organisms (the stalked ciliates such as *Vorticella* sp., for example) or they can colonize like the sedentary *Carchesium* sp. As the principal hunters and grazers of the microbial world, protozoa play a key role in maintaining the balance of bacterial, algal, and other microbial life. Protozoa are microscopic and get their name because they employ the same type of feeding strategy as animals. The animal-like protozoans differ from animals in that they are unicellular and do not have specialized tissues, organs, or organ systems that carry out life functions. Most are harmless, but some are parasitic. Some forms have 2 life stages: active **trophozoites** (capable of feeding) and dormant **cysts** (Spellman 2000).

✓ **Important Point**: Although they are efficient hunters and grazers—feeding on bacteria, eating other protozoa and bits of material that has come off of other living things (organic matter)—protozoans are themselves an important food source for larger creatures and the basis of many food chains.

As mentioned, as unicellular eukaryotes, protozoa cannot be easily defined because they are diverse and, in most cases, only distantly related to each other. Also, again, protozoa are distinguished from bacteria by their eukaryotic nature and by their usually larger size. Protozoa are distinguished from algae because protozoa obtain energy and nutrients by taking in organic molecules, detritus, or other protozoans rather than from photosynthesis. Each protozoan is a complete organism and contains the facilities for performing all the body functions for which vertebrates have many organ systems.

Like bacteria, protozoa depend upon environmental conditions (the protozoan quickly responds to changing physical and chemical characteristics of the environment), reproduction, and availability of food for their existence. Relatively large microorganisms, protozoans range in size from 4 microns to about 500 microns. They can both consume bacteria (limit growth) and feed on organic matter (degrade waste) (Spellman 2000).

Interest in types of protozoa is high among water treatment practitioners because certain types of protozoans can cause disease. In the United States, the most important of the pathogenic parasitic protozoans is *Giardia lamblia*, which causes a disease known as **giardiasis**. Two other parasitic protozoans that carry waterborne disease are *Entamoeba histolytica* (amoebic dysentery) and *Cryptosporida* (cryptosporidosis) (Spellman 2000).

Protozoa are divided into 4 groups based on their method of motility as shown in Table 6.2.

- **Mastigophora**—these protozoans are mostly unicellular, lack specific shape (have an extremely flexible plasma membrane that allows for movement of cytoplasm), and possess whip-like structures called flagella. The flagella—which use a whip-like motion to move in a relatively straight path, or create currents that spin the organism through fluids—are used for locomotion, as sense receptors, and to attract food.

  These organisms are common in both fresh and marine waters. The group's subdivisions include the *phytomastigophorea* class, most of which contain chlorophyll and are thus plant-like. A characteristic species of *phytomastigophorea* is the *Euglena* sp., often associated with high or increasing levels of nitrogen and phosphate in the wastewater treatment process.

  ✓ **Interesting Point**: The mastigophora trypanosomes require 2 hosts, one a mammal, to complete their life cycle. They cause the disease African sleeping sickness, Chagas disease, and leishmaniasis. Trichonymphs are symbionts inside the intestines of termites.

Table 6.2.   Classification of Protozoans

Group	Common Name	Movement	Reproduction
Mastigophora	Flagellates	Flagella	Asexual
Ciliophora	Ciliates	Cili	Asexual by transverse fission Sexual by conjugation
Sarcodina	Amoebas	Pseudophodia	Asexual and sexual
Sporozoa	Sporozoans	Nonmotile	Asexual and sexual

- **Ciliophora** (Ciliates)—the ciliates are the most advanced and structurally complex of all protozoans. They are heterotrophic and use multiple small cilia for locomotion. The *Paramecium* is probably the most commonly studied ciliate in basic biology classes. Movement and food-getting is accomplished with short hairlike structures called cilia that are present in at least 1 stage of the organism's life cycle. Three groups of ciliates exist: free-swimmers, crawlers, and stalked. The majority are free-living. They are usually solitary, but some are colonial and others are sessile (attached and not free-moving). They are unique among protozoa in having 2 kinds of nuclei: a micronucleus and a macronucleus. The micronucleus is concerned with sexual reproduction. The macronucleus is involved with metabolism and the production of RNA for cell growth and function.

  ✓ **Interesting Point**: "To increase strength of the cell boundary, ciliates have a *pellicle*, a sort of tougher membrane that still allows them to change shape" (Farabee 2007).

  The ciliate pellicle may also act as thick armor. In other species, the pellicle may be very thin. The cilia are short and usually arranged in rows. Their structure is comparable to flagella except that cilia are shorter. Cilia may cover the surface of the animal or may be restricted to banded regions.

  Like many freshwater protozoans, ciliates are hypotonic; however, removal of water crossing the cell membrane by osmosis is a problem. Therefore, one commonly employed mechanism is a contractile vacuole. Water is collected into the central ring of the vacuole and actively transported from the cell.

- **Sarcodina** (pseudopods, "false feet")—members of this group have fewer organelles and are simpler in structure than the ciliates and flagellates. Sarcodina move about by the formation of flowing protoplasmic projections called **pseudopodia**. The formation of pseudopodia is commonly referred to as **amoeboid movement**. The **amoebae** are well known for this mode of action (see Figure 6.4).

  The pseudopodia not only provide a means of locomotion, but also serve as a means of feeding; this is accomplished when the organism puts out the pseudopodium to enclose the food. Most amoebas feed on algae, bacteria, protozoa, and rotifers. Several species in the Sarcodina group, including some species of amoebas, cover themselves with protective shell-like coverings call *tests*. These tests (made of silica) are stippled with many small and large openings through which water can flow in and out, and through which the pseudopods protrude.

  ✓ **Interesting Point**: "Pseudopodia are used by many cells and are not fixed structures like flagella but rather are associated with actin near the moving edge of the cytoplasm" (Farabee 2007).

- **Sporozoans**—these protozoans are obligatory intracellular parasites: they must spend at least part if not all of their life cycle in a host animal. They have no special structures used for locomotion. The life cycle often involves more than one host, such as when plasmodium infects both mosquitoes and humans, causing human malaria.

  ✓ **Interesting Point**: The plasmodium includes the malaria parasites transmitted by Anopheles mosquitoes.

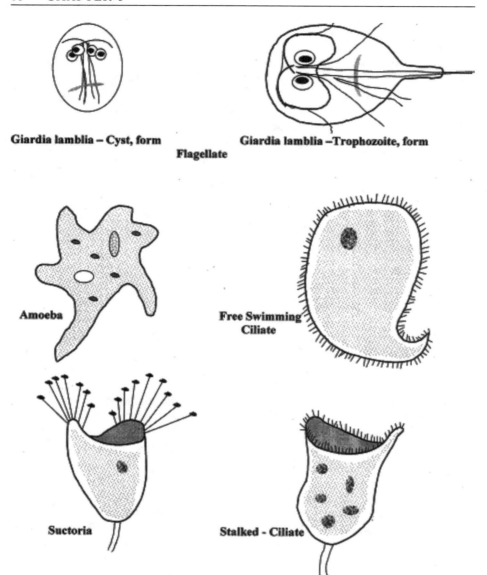

Giardia lamblia – Cyst, form
Flagellate

Giardia lamblia –Trophozoite, form

Amoeba

Free Swimming
Ciliate

Suctoria

Stalked - Ciliate

Figure 6.4.   Amoebae and other protozoans.

ALGAE

The protists that perform photosynthesis are called **algae**. Algae can be both a nuisance and an ally. Many ponds, lakes, rivers, streams, and bays (e.g., Chesapeake Bay) in the United States (and elsewhere) are undergoing **eutrophication**, the enrichment of an environment with inorganic substances (phosphorous and nitrogen). When eutrophication occurs, and filamentous algae like *Caldophora* break loose in a pond, lake, stream, or river and wash ashore, the algae's stinking, noxious presence is clearly evident. Algae are also allies in many wastewater treatment operations. They can be valuable in long-term oxidation ponds where they aid in the purification process by producing oxygen.

Before discussing the specifics and different types of algae, it is important to be familiar with algal terminology:

- **Algae**—large and diverse assemblages of eukaryotic organisms that lack roots, stems, and leaves but have chlorophyll and other pigments for carrying out oxygen-producing photosynthesis
- **Algology** or **Phycology**—the study of algae
- **Antheridium**—special male reproductive structures where sperm are produced
- **Aplanospore**—nonmotile spores produced by sporangia
- **Benthic**—algae attached and living on the bottom of a body of water
- **Binary fission**—nuclear division followed by division of the cytoplasm
- **Chloroplasts**—packets that contain *chlorophyll a* and other pigments
- **Chrysolaminarin**—the carbohydrate reserve in organisms of division **Chrysophyta**
- **Diatoms**—photosynthetic, circular, or oblong chrysophyte cells
- **Dinoflagellates**—unicellular, photosynthetic protistan algae
- **Epitheca**—the larger part of the frustule (diatoms)
- **Euglenids**—contain chlorophylls **a** and **b** in their chloroplasts; representative genus is *Euglena*
- **Fragmentation**—a type of asexual algal reproduction in which the thallus breaks up and each fragmented part grows to form a new thallus
- **Frustule**—the distinctive 2-piece wall of silica in diatoms
- **Hypotheca**—the small part of the frustule (diatoms)
- **Neustonic**—algae that live at the water-atmosphere interface
- **Oogonia**—vegetative cells that function as female sexual structures in the algal reproductive system
- **Pellicle**—a *Euglena* structure that allows for turning and flexing of the cell
- **Phytoplankton**—made up of algae and small plants
- **Plankton**—free-floating, mostly microscopic aquatic organisms
- **Planktonic**—algae suspended in water as opposed to attached and living on the bottom (benthic)
- **Prototothecosis**—a disease in humans and animals caused by the green algae, *Prototheca moriformis*
- **Thallus**—the vegetative body of algae

Algae are autotrophic, contain the green pigment chlorophyll, and are a form of aquatic plant. Algae differ from bacteria and fungi in their ability to carry out photosynthesis—the biochemical process requiring sunlight, carbon dioxide, and raw mineral nutrients. Photosynthesis takes place in the chloroplasts. The chloroplasts are usually distinct and visible. They vary in size, shape, distribution, and numbers. In some algal types the chloroplast may occupy most of the cell space. They usually grow near the surface of water because light cannot penetrate very far through water. Although larger species (multicellular forms like marine kelp) can easily be seen by the unaided eye, many of them are microscopic. Algal cells may be nonmotile, motile by one or more flagella, or exhibit gliding **motility** as in diatoms. They occur most commonly in water (fresh and polluted water, as well as in salt water), in which they may be suspended (planktonic) phytoplanktons or attached and living on the bottom (benthic). A few algae live at the water-atmosphere interface and are termed *neustonic*. Within the fresh and saltwater environments, algae are important primary producers (the start of the food chain for other organisms). During their growth phase, they are important oxygen-generating organisms and constitute a significant portion of the plankton in water.

According to the 5-kingdom system of Whittaker, the algae belong to 7 divisions distributed between 2 different kingdoms. Although 7 divisions of algae occur, only 5 divisions are discussed in this text:

- **Chlorophyta**—green algae
- **Euglenophyta**—euglenids
- **Chrysophyta**—golden-brown algae, diatoms
- **Phaeophyta**—brown algae
- **Pyrrophyta**—dinoflagellates

The primary classification of algae is based on cellular properties. Several characteristics are used to classify algae, including: (1) cellular organization and cell wall structure; (2) the nature of chlorophyll(s) present; (3) the type of motility, if any; (4) the carbon polymers that are produced and stored; and (5) the reproductive structures and methods. Table 6.3 summarizes the properties of the 5 divisions discussed in this text.

Algae show considerable diversity in the chemistry and structure of their cells. Some algal cell walls are thin, rigid structures usually composed of cellulose modified by the addition of other polysaccharides. In other algae, the cell wall is strengthened by the deposition of calcium carbonate. Other forms have chitin present in the cell wall. Complicating the classification of algal organisms are the euglenids, which lack cell walls. In diatoms the cell wall is composed of silica. The frustules (shells) of diatoms exhibit extreme resistance to decay and remain intact for long periods of time, as the fossil records indicate.

The principal feature used to distinguish algae from other microorganisms (for example, fungi) is the presence of chlorophyll and other photosynthetic pigments in algae. All algae contain chlorophyll a. Some, however, contain other types of chlorophylls. The presence of these additional chlorophylls is characteristic of a particular algal group. In addition to chlorophyll, other pigments encountered in algae include fucoxanthin (brown), xanthophylls (yellow), carotenes (orange), phycocyanin (blue), and phycoerythrin (red).

Many algae have flagella (a threadlike appendage). As mentioned, the flagella are locomotor organelles that may be the single polar or multiple polar types. The *Euglena* is a simple flagellate form with a single polar flagellum. Chlorophyta have either 2 or 4 polar flagella. Dinoflagellates have 2 flagella of different lengths. In some cases, algae

Table 6.3. Comparative Summary of Algal Characteristics

Algal Group	Common Name	Structure	Pigments	Carbon Reserve	Motility	Reproduction
Chlorophyta	Green algae	Unicellular to Multicellular	Chlorophylls a and b, carotenes, xanthophylls	Starch, oils	Most are nonmotile	Asexual and sexual
Euglenophyta	Euglenoids	Unicellular	Chlorophylls a and b, carotenes, xanthophylls	Fats	Motile	Asexual
Chrysophyta	Golden brown algae, diatoms	Multicellular	Chlorophylls a and b, special carotenoids, xanthophylls	Oils	Gliding by diatoms; others by flagella	Asexual and sexual
Phaeophyta	Brown algae	Unicellular	Chlorophylls a and b, carotenoids xanthophylls	Fats	Motile	Asexual and sexual
Pyrrophyta	Dinoflagellated	Unicellular	Chlorophylls a and b, Carotenes, xanthophylls	Starch	Motile	Asexual; sexual rare

are nonmotile until they form motile gametes (a haploid cell or nucleus) during sexual reproduction. Diatoms do not have flagella, but have gliding motility.

Algae can be either autotrophic or heterotrophic. Most are photoautotrophic; they require only carbon dioxide and light as their principal source of energy and carbon. In the presence of light, algae carry out oxygen-evolving photosynthesis; in the absence of light, algae use oxygen. Chlorophyll and other pigments are used to absorb light energy for photosynthetic cell maintenance and reproduction. One of the key characteristics used in the classification of algal groups is the nature of the reserve polymer synthesized as a result of utilizing carbon dioxide present in water.

Algae may reproduce either asexually or sexually. Three types of asexual reproduction occur: binary fission, spores, and fragmentation. In some unicellular algae, binary fission occurs where the division of the cytoplasm forms new individuals like the parent cell following nuclear division. Some algae reproduce through spores. These spores are unicellular and germinate without fusing with other cells. In fragmentation, the thallus breaks up and each fragment grows to form a new thallus.

Sexual reproduction can involve a union of cells, where eggs are formed within vegetative cells called *oogonia* (which function as female structures) and sperm are produced in a male reproductive organ called *antheridia*. Algal reproduction can also occur through a reduction of chromosome number and/or the union of nuclei.

Characteristics of Algal Divisions

- **Chlorophyta** (green algae)—the majority of algae found in ponds belong to this group; they also can be found in salt water and soil. Several thousand species of green algae are known today. Many are unicellular; others are multicellular filaments or aggregated colonies. The green algae have chlorophylls a and b, along with specific carotenoids, and they store carbohydrates as starch. Few green algae are found at depths greater than 7 to 10 meters, largely because sunlight does not penetrate to that depth. Some species have a holdfast structure that anchors them to the bottom of the pond and to other submerged inanimate objects. Green algae reproduce by both sexual and asexual means. Multicellular green algae have some division of labor, producing various reproductive cells and structures.
- **Euglenophyta** (euglenids)—are a small group of unicellular microorganisms that have a combination of animal and plant properties. Euglenids lack a cell wall, possess a gullet, have the ability to ingest food, have the ability to assimilate organic substances, and, in some species, are absent of chloroplasts. They occur in fresh, brackish, and salt waters, and on moist soils. A typical *Euglena* cell is elongated and bound by a plasma membrane; the absence of a cell wall makes them very flexible in movement. Inside the plasma membrane is a structure called the pellicle that gives the organisms a definite form and allows the cell to turn and flex. Euglenids that are photosynthetic contain chlorophylls a and b, and they always have a red eyespot (**stigma**) that is sensitive to light (photoreceptive). Some euglenids move about by means of flagellum; others move about by means of contracting and expanding motions. The characteristic food supply for euglenids is a lipopolysaccharide. Reproduction in euglenids is by simple cell division.

✓   **Interesting Point**: Some autotrophic species of *Euglena* become heterotrophic when light levels are low.

- **Chrysophyta** (golden-brown algae)—the Chrysophycophyta division of the phylum Chrysophyta is quite large—several thousand diversified members. Chrysophycophyta differ from green algae and euglenids in that: (1) chlorophylls a and c are present; (2) fucoxanthin, a brownish pigment, is present; and (3) they store food in the form of oils and leucosin, a polysaccharide. The combination of yellow pigments, fucoxanthin, and chlorophylls causes most of these algae to appear golden-brown. The Chrysophycophyta is also diversified in cell wall chemistry and flagellation. The division is divided into 3 major classes: golden-brown, yellow-brown algae, and diatom.

  Some Chrysophyta lack cell walls; others have intricately patterned coverings external to the plasma membrane, such as walls, plates, and scales. The diatoms are the only group that has hard cell walls of pectin, cellulose, or silicon, constructed in 2 halves (the epitheca and the hypotheca) called a frustule. Two anteriorly attached flagella are common among Chrysophyta; others have no flagella.

  Most Chrysophyta are unicellular or colonial. Asexual cell division is the usual method of reproduction in diatoms; other forms of Chrysophyta can reproduce sexually.

  Diatoms have direct significance for humans. Because they make up most of the phytoplankton of the cooler ocean parts, they are the ultimate source of food for fish.

  Water and wastewater operators understand the importance of diatoms, which function as indicators of industrial water pollution. As water quality indicators, their specific tolerances to environmental parameters such as pH, nutrients, nitrogen, concentration of salts and temperature have been compiled.

  ✓   **Interesting Point**: Diatoms secrete a silicon dioxide shell (frustule) that forms the fossil deposits known as diatomaceous earth, used in filters, and as abrasives in polishing compounds.

- **Phaeophyta** (brown algae)—with the exception of a few freshwater species, all algal species of this division exist in marine environments as seaweed. They are a highly specialized group, consisting of multicellular organisms that are sessile. These algae contain essentially the same pigments seen in the golden-brown algae, but they appear brown because the predominance of fucoxanthin has a masking effect on the other pigments. Brown algal cells store food as the carbohydrate laminarin and some lipids. Brown algae reproduce asexually. They are used in foods, animal feeds, and fertilizers, and as source for alginate, a chemical emulsifier added to ice cream, salad dressing, and candy.

- **Pyrrophyta** (dinoflagellates)—the principal members of this division are the dinoflagellates. The dinoflagellates comprise a diverse group of biflagellated and nonflagellated unicellular, eukaryotic organisms. The dinoflagellates occupy a variety of aquatic environments with the majority living in marine habitats. Most of these organisms have a heavy cell wall composed of cellulose-containing plates. They store food as starch, fats, and oils. These algae have chlorophylls a and c and several xanthophylls. The most common form of reproduction in dinoflagellates is by cell division, but sexual reproduction has also been observed.

✓   **Interesting Point**: Cell division in dinoflagellates differs from most protistans, with chromosomes attaching to the nuclear envelope and being pulled apart as the nuclear envelope stretches. During cell division in most other eukaryotes, the nuclear envelope dissolves.

SLIME AND WATER MOLDS

Slime mold, most commonly found in a forest, is a heterotrophic organism that was once regarded as a fungus but later classified with the Protista. They have very complex life cycles involving multiple forms and stages. In their visible, aggregate states, they look like blobs. They may be red, yellow, brown, bright orange, black, white, or blue. They spend part of their life as single-celled forms, but can aggregate to form multicellular forms. Slime molds eat decaying vegetation, bacteria, fungi, and even other slime molds.

Water molds (Oomycota) are decomposers found in freshwater aquatic environments. They are known as downy mildews and white rusts. Downy mildew caused the Great Potato Famine in Ireland in 1845–1849.

✓   **Interesting Point**: Slime molds act like giant amoebas, creeping slowly along and engulfing food particles along the way.

Fungi

Fungi (singular fungus) constitute an extremely important and interesting group of eukaryotic, aerobic microbes ranging from the unicellular yeasts to the extensively mycelial molds. Fungi first evolved in water but made the transition to land through the development of specialized structures that prevented their drying out. Not considered plants, they are a distinctive life-form of great practical and ecological importance. Fungi are important because, like bacteria, they metabolize dissolved organic matter; they are the principal organisms responsible for the decomposition of carbon in the biosphere. Fungi, unlike bacteria, can grow in low moisture areas and in low pH solutions, which aids them in the breakdown of organic matter.

Before discussing specifics and the different types of fungi, it is important to be familiar with fungal terminology:

- **Budding**—process by which yeasts reproduce.
- **Blastospore** or **bud**—spores formed by budding.
- **Conidia**—asexual spores that form on specialized hyphae called **conidiophores**; large conidia are called **macroconidia** and small conidia are called **microconidia.**

- **Hypha** (pl. **hyphae**)—a tubular cell that grows from the tip and may form many branches.
- **Mycelium**—consists of many-branched hypha and can become large enough to be seen with the naked eye.
- **Nonseptate** or **aseptate**—when cross walls are not present.
- **Septate hyphae**—when a filament has crosswalls.
- **Sexual spores**—in the fungi division Amastigomycota, 4 subdivisions are separated on the basis of type of sexual reproductive spores present:

  1. **Zygomycotina**—consists of nonseptate hyphae and zygospores. Zygospores are formed by the union of nuclear material from the hyphae of 2 different strains.
  2. **Ascomycotina**—fungi in this group are commonly referred to as the ascomycetes. They are also called **sac fungi**. They all have septate hyphae. **Ascospores** are the characteristic sexual reproductive spores and are produced in sacs called **asci** (ascus, singular). The mildews and **Penicillium** with asci in long fruiting bodies belong to this group.
  3. **Basidiomycotina**—consists of mushrooms, puffballs, smuts, rust, and shelf fungi (found on dead trees). The sexual spores of this class are known as **basidiospores**, which are produced on the club-shaped **basidia**.
  4. **Deutermycotina**—consists of only one group, the **Deuteromycetes**. Members of this class are referred to as the **fungi imperfecti** and include all the fungi that lack sexual means of reproduction.

- **Sporangiospores**—spores that form within a sac called a **sporanguim**; the sporangia are attached to stalks called **sporangiophores.**
- **Spore**—reproductive stage of the fungi.

Fungi comprise a large group of organisms that include such diverse forms as molds, mushrooms, puffballs, and yeasts. Because they lack chlorophyll (and thus are not considered plants) they must get nutrition from organic substances. They are either **parasites**, existing in or on animals or plants, or more commonly are **saprophytes**, obtaining their food from dead organic matter. Fungi also are important crop parasites, causing loss of food plants, spoilage of food and some infectious diseases. Fungi are classified in their own kingdom but the main groups are called divisions rather than phyla. The study of fungi is called **mycology**.

✓ **Interesting Point**: Fungi range in size from the single-celled organism we know as yeast to the largest known living organism on Earth—a 3.5-mile-wide mushroom dubbed "the humongous fungus," which covers some 2,000+ acres in Oregon's Malheur National Forest.

McKinney, in *Microbiology for Sanitary Engineers*, complains that the study of mycology has been directed solely toward classification of fungi and not toward the actual biochemistry involved with fungi. McKinney goes on to point out that for those involved in the sanitary field it is important to recognize the "sanitary importance of fungi . . . and other steps will follow" (1962, 40). For students of environmental science, understanding the role of fungi as it relates to the water purification process is important. Environmental practitioners need knowledge and understanding of the organism's ability to function and exist under extreme conditions, which make fungi important elements in biological wastestream treatment processes and in the degradation that takes place during waste-composting processes.

Fungi may be unicellular or filamentous. They are large, 5 to 10 microns wide, and can be identified by a microscope. The distinguishing characteristics of the group as a whole are that they: (1) are nonphotosynthetic, (2) lack tissue differentiation, (3) have cell walls of polysaccharides (chitin), and (4) propagate by spores (sexual or asexual).

Fungi are divided into 5 classes:

- **Myxomycetes**, or slime fungi
- **Phycomycetes**, or aquatic fungi (algae)
- **Ascomycetes**, or sac fungi
- **Basidiomycetes**, or rusts, smuts, and mushrooms
- **Fungi imperfecti**, or miscellaneous fungi

✔   **Interesting Point**: Although fungi are limited to only 5 classes, more than 80,000 known species exist.

Fungi differ from bacteria in several ways, including in their size, structural development, methods of reproduction, and cellular organization. They differ from bacteria in another significant way as well: their biochemical reactions (unlike the bacteria) are not important for classification; instead, their structure is used to identify them. Fungi can be examined directly, or suspended in liquid, stained, dried, and observed under microscopic examination where they can be identified by appearance (color, texture, and diffusion of pigment) or their mycelia.

One of the tools available to environmental science students and specialists for use in the fungal identification process is the distinctive terminology used in mycology. Fungi go through several phases in their life cycle; their structural characteristics change with each new phase. Become familiar with the following listed and defined terms. As a further aid in learning how to identify fungi, relate the defined terms to their diagrammatic representation (Figure 6.5).

Fungi can be grown and studied by cultural methods. However, when culturing fungi, use culture media that limits the growth of other microbial types—controlling bacterial growth is of particular importance. This can be accomplished by using special agar (culture media) that depresses pH of the culture medium (usually Sabouraud glucose or maltose agar) to prevent the growth of bacteria. Antibiotics can also be added to the agar that will prevent bacterial growth.

✔   **Interesting Point**: Fungi can be found in rising bread, moldy bread, old food in the refrigerator, and on forest floors.

As part of their reproductive cycle, fungi produce very small spores that are easily suspended in air and widely dispersed by the wind. Insects and other animals also

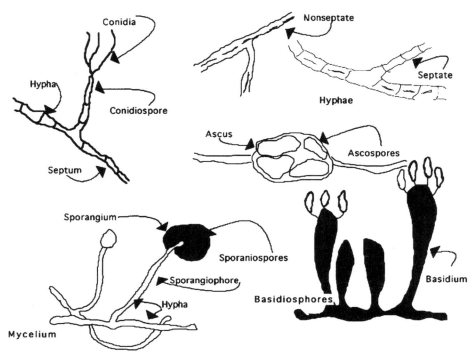

Figure 6.5.   Nomenclature of fungi. Adapted from McKinney (1962, 36).

spread fungal spores. The color, shape, and size of spores are useful in the identification of fungal species.

Reproduction in fungi can be either sexual or asexual. The union of compatible nuclei accomplishes sexual reproduction. Most fungi form specialized asexual and/or sexual spore-bearing structures (fruiting bodies). Some fungal species are self-fertilizing and other species require outcrossing between different but compatible vegetative thalluses (mycelia).

Most fungi are asexual. Asexual spores are often brightly pigmented and give their colony a characteristic color (green, red, brown, back, blue—the blue spores of *Penicillium roquefort* are found in blue or Roquefort cheese).

✓   **Interesting Point**: Fungi usually reproduce without sex. Single-celled yeasts reproduce asexually by budding. A single yeast cell can produce up to 24 offspring.

Asexual reproduction is accomplished in several ways:

- Vegetative cells may bud to produce new organisms. This is very common in the yeasts.
- A parent cell can divide into 2 daughter cells.
- The most common method of asexual reproduction is the production of spores.

Several types of asexual spores are common:

- If a hypha separates to form cells that behave as spores, they are called **arthrospores**.
- If a thick wall encloses the cells before separation, they are called **chlamydospores**.
- If budding produces the spores, they are called **blastospores**.
- If the spores develop within sporangia (sac), they are called **sporangiospores**.
- If the spores are produced at the sides or tips of the hypha, they are called **conidiospores**.

Fungi are found wherever organic material is available. They prefer moist habitats and grow best in the dark. Most fungi can best be described as grazers, but a few are active hunters. Most fungi are saprophytes, acquiring their nutrients from dead organic matter, gained when the fungi secrete hydrolytic enzymes, which digest external substrates. They are able to use dead organic matter as a source of carbon and energy. Most fungi use glucose and maltose (carbohydrates) and nitrogenous compounds to synthesize their own proteins and other needed materials. Knowing from what materials fungi synthesize their own protein and other needed materials in comparison to what bacteria are able to synthesize is important to those who work in the environmental disciplines for understanding the growth requirements of the different microorganisms.

✓   **Interesting Point**: Some fungi produce a sticky substance on their hyphae, which then act like flypaper, trapping passing prey.

Plants

The plant kingdom ranks second in importance only to the animal kingdom (at least from the human point of view). The importance of plants and plant communities to humans and their environment cannot be overstated. Some of the important things plants provide are listed below:

- **Aesthetics**—plants add to the beauty of the places we live.
- **Medicine**—80% of all medicinal drugs originate in wild plants.
- **Food**—90% of the world's food comes from only 20 plant species.
- **Industrial products**—plants are very important for the goods they provide. For example, plant fibers provide clothing; wood is used to build homes; and some important fuel chemicals come from plants, such as ethanol from corn and soy diesel from soybeans.
- **Recreation**—plants form the basis for many important recreational activities, including fishing, nature observation, hiking, and hunting.
- **Air quality**—the oxygen in the air we breathe comes form the photosynthesis of plants.
- **Water quality**—plants aid in maintaining healthy watersheds, streams, and lakes by holding soil in place, controlling stream flows, and filtering sediments from water.

Table 6.4.   Major Differences between Plants and Animals

Plants	Animals
Plants contain chlorophyll and can make their own food.	Animals cannot make their own food and are dependent on plants and other animals for food.
Plants give off oxygen and take in carbon dioxide given off by animals.	Animals give off carbon dioxide which plants need to make food and take in oxygen which they need to breathe.
Plants generally are rooted in one place and do not move on their own.	Most animals have the ability to move fairly freely.
Plants have either no or very basic ability to sense.	Animals have a much more highly developed sensory and nervous system.

- **Erosion control**—plant cover helps to prevent wind or water erosion of the top layer of soil that we depend on.
- **Climate**—regional climates are impacted by the amount and type of plant cover.
- **Fish and wildlife habitat**—plants provide the necessary habitat for wildlife and fish populations.
- **Ecosystem**—every plant species serves an important role or purpose in their community.

Though both are important kingdoms of living things, plants and animals differ in many important aspects. Some of these differences are summarized in Table 6.4.

DEFINITION OF KEY TERMS

Before discussing the basic specifics of plants, it is important to first define a few key plant terms:

- **Apical meristem**—consists of meristematic cells located at tip (apex) of a root or shoot.
- **Cambium**—the lateral meristem in plants.
- **Chloroplasts**—disk-like organelles with a double membrane found in eukaryotic plant cells.
- **Companion cells**—specialized cells in the phloem that load sugars into the sieve elements.
- **Cotyledons**—leaf-like structure (sometimes referred to as "seed leaf") that is present in the seeds of flowering plants.
- **Dicot**—one of the 2 main types of flowering plants; characterized by having 2 cotyledons
- **Diploid**—having 2 of each kind of chromosome (2n).
- **Guard cells**—specialized epidermal cells that flank stomata and whose opening and closing regulate gas exchange and water loss.

- **Haploid**—having only a single set of chromosomes (n).
- **Meristem**—group of plant cells that can divide indefinitely, provides new cells for the plant.
- **Monocots**—one of 2 main types of flowering plants; characterized by having a single cotyledon.
- **Periderm**—a layer of plant tissue derived from the cork cambium, and then secondary tissue, replacing the epidermis.
- **Phloem**—complex vascular tissue that transports carbohydrates throughout the plant.
- **Sieve cells**—conducting cells in the phloem of vascular plants.
- **Stomata**—pores on the underside of leaves that can be opened or closed to control gas exchange and water loss.
- **Thallus**—main plant body, not differentiated into a stem or leaves.
- **Tropism**—plant behavior; controlling the direction of plant growth.
- **Vascular tissue**—tissues found in the bodies of vascular plants that transport water, nutrients, and carbohydrates. The 2 major kinds are xylem and phloem.
- **Xylem**—vascular tissue of plants that transports water and dissolved minerals from the roots upwards to other parts of plant. Xylem often also provides mechanical support against gravity.

Although not typically acknowledged, plants are as intricate and complicated as animals. Plants evolved from photosynthetic protists and are characterized by photosynthetic nutrition, cell walls made from cellulose and other polysaccharides, lack of mobility and a characteristic life cycle involving an alternation of generations. The phyla/division of plants and examples are listed in Table 6.5.

THE PLANT CELL

The cell was covered earlier, but a brief summary of plant cells is provided here:

- *Plants have all the organelles animal cells have* (i.e., nucleus, ribosomes, mitochondria, endoplasmic reticulum, Golgi apparatus, etc.).

Table 6.5.   The Main Phyla/Division of Plants

Phylum/Division	Examples
Bryophyta	Mosses, liverworts, and hornworts
Coniferophyta	Conifers such as redwoods, pines, and firs
Cycadophyta	Cycads, sago palms
Gnetophyta	Shrub trees and vines
Ginkophyta	*Ginkgo* is the only genus
Lycophyta	Lycopods (look like mosses)
Pterophyta	Ferns and tree-ferns
Anthophyta	Flowering plants including oak, corn, maize, and herbs

- *Plants have chloroplasts.* Chloroplasts are special organelles that contain chlorophyll and allow plants to carry out photosynthesis.
- *Plant cells can sometimes have large vacuoles for storage.*
- *Plant cells are surrounded by a rigid cell wall made of cellulose,* which provide support, in addition to the cell membrane that surrounds animal cells (Spellman 2000).

VASCULAR PLANTS

Vascular plants, also called **Tracheophytes**, have special vascular tissue for the transport of necessary liquids and minerals over long distances. Vascular tissues are composed of specialized cells that create "tubes" through which materials can flow throughout the plant body. These vessels are continuous throughout the plant, allowing for the efficient and controlled distribution of water and nutrients. In addition to this transport function, vascular tissues also support the plant. The 2 types of vascular tissue are xylem and phloem:

- **Xylem**—consists of a tube or a tunnel (pipeline) in which water and minerals are transported throughout the plant to leaves for photosynthesis. In addition to distributing nutrients, xylem (wood) provides structural support. After a time, the xylem at the center of older trees ceases to function in transport and takes on a supportive role only.
- **Phloem tissue**—consists of cells called *sieve tubes* and *companion cells.* Phloem tissue moves dissolved sugars (carbohydrates), amino acids and other produces of photosynthesis from the leaves to other regions of the plant.

The 2 most important Tracheophytes are gymnosperms (gymno = naked; sperma = seed) and angiosperms (angio = vessel, receptacle, container):

- **Gymnosperms**—the plants we recognize as gymnosperms represent the sporophyte generation (i.e., the spore-producing phase in the life cycle of a plant that exhibits alternation of generation). Gymnosperms were the first tracheophytes to use seeds for reproduction. The seeds develop in protective structures called cones. A gymnosperm contains some cones that are female and some that are male. Female cones produce spores that, after fertilization, become eggs enclosed in seeds that fall to the ground. Male cones produce pollen, which is taken by the wind and fertilizes female eggs by that means. Unlike flowering plants, the gymnosperm does not form true flowers or fruits. Coniferous tress such as firs and pines are good examples of gymnosperms.
- **Angiosperms**—the flowering plants, are the most highly evolved plants and the most dominant in present times. They have stems, roots, and leaves. Unlike gymnosperms such as conifers and cycads, angiosperm's seeds are found in a flower. Angiosperm eggs are fertilized and develop into a seed in an ovary that is usually in a flower.

NatureWorks (2007) points out that there are 2 types of angiosperms: monocots and dicots:

- **Monocots**—these angiosperms start with 1 seed-leaf (cotyledon); thus, their name, which is derived from the presence of a single cotyledon during embryonic development. Monocots include grasses, grains, and other narrow-leaved angiosperms. The main veins of their leaves are usually parallel and unbranched, the flower parts occur in multiples of 3, and a fibrous root system is present. Monocots include orchids, lilies, irises, palms, grasses, and wheat, corn and oats.
- **Dicots**—angiosperms in this group grow 2 seed-leaves (2 cotyledons). Most plants are dicots and include maples, oaks, elms, sunflowers, and roses. Their leaves usually have a single main vein or 3 of more branched veins that spread out from the base of the leaf.

LEAVES

The principal function of leaves is to absorb sunlight for the manufacturing of plant sugars in photosynthesis. The leaves' broad, flattened surfaces gather energy from sunlight, while apertures on their undersides bring in carbon dioxide and release oxygen. Leaves develop as a flattened surface in order to present a large area for efficient absorption of light energy. On its 2 exteriors, the leaf has layers of epidermal cells that secrete a waxy, nearly impermeable cuticle (chitin) to protect against water loss (dehydration) and fungal or bacterial attack. Gases diffuse in or out of the leaf through **stomata**, small openings on the underside of the leaf. The opening or closing of the stomata occurs through the swelling or relaxing of **guard cells**. If the plant wants to limit the diffusion of gases and the transpiration of water, the guard cells swell together and close the stomata. Leaf thickness is kept to a minimum so that gases that enter the leaf can diffuse easily throughout the leaf cells.

CHLOROPHYLL/CHLOROPLAST

The green pigment in leaves is **chlorophyll**. Chlorophyll absorbs red and blue light from the sunlight that falls on leaves. Therefore, the light reflected by the leaves is diminished in red and blue and appears green. The molecules of chlorophyll are large. They are not soluble in the aqueous solution that fills plant cells. Instead, they are attached to the membranes of disk-like structures, called **chloroplasts**, inside the cells. Chloroplasts are the site of photosynthesis, the process in which light energy is converted to chemical energy. In chloroplasts, the light absorbed by chlorophyll supplies the energy used by plants to transform carbon dioxide and water into oxygen and carbohydrates.

Chlorophyll is not a very stable compound; bright sunlight causes it to decompose. To maintain the amount of chlorophyll in their leaves, plants continuously synthesize it. The synthesis of chlorophyll in plants requires sunlight and warm temperatures. Therefore, during summer chlorophyll is continuously broken down and regenerated in the leaves of trees.

PHOTOSYNTHESIS

Because our quality of life, and indeed our very existence, depends on photosynthesis, it is essential to understand it. In photosynthesis, plants (and other photosynthetic autotrophs) use the energy from sunlight to create the carbohydrates necessary for cell respiration. More specifically, plants take water and carbon dioxide and transform them into glucose and oxygen:

$$6CO_2 + 6H_2O + \text{light energy} \rightarrow C_6H_{12}O_6 + 6O_2$$

This general equation of photosynthesis represents the combined effects of 2 different stages. The first stage is called the light reaction and the second stage is called the dark reaction. The **light reaction** is the photosynthetic process in which solar energy is harvested and transferred into the chemical bonds of ATP. It can only occur in light. The **dark reaction** is the process in which food (sugar) molecules are formed from carbon dioxide from the atmosphere with the use of ATP. This process can occur in the dark as long as ATP is present.

✓   **Interesting Point**: Charles Darwin was the first to discuss how plants respond to light. He found that new shoots of grasses bend toward the light because the cells on the dark side grow faster than the lighted side.

ROOTS

"**Roots** absorb nutrients and water, anchor the plant in the soil, provide support for the stem, and store food. They are usually below ground and lack nodes, shoots, and leaves. There are 2 major types of root systems in plants. Taproot systems have a stout main root with a limited number of side-branching roots. Examples of taproot system plants are nut trees, carrots, radishes, parsnips, and dandelions. Taproots make transplanting difficult. The second type of root system, fibrous, has many branched roots. Examples of fibrous root plants are most grasses, marigolds and beans" (Master Gardener 2006). Radiating from the roots is a system of root hairs, which vastly increase the absorptive surface area of the roots. Roots also anchor the plant in the soil.

GROWTH IN VASCULAR PLANTS

Vascular plants undergo 2 kinds of growth (growth is primarily restricted to meristems), primary and secondary growth. **Primary growth** occurs relatively close to the tips of roots and stems. It is initiated by apical meristems and it is primarily involved in the extension of the plant body. The tissues that arise during primary growth are called primary tissues and the plant body composed of these tissues is called the primary plant body. Most primitive vascular plants are entirely made up of primary tissues. **Secondary growth** occurs in some plants; secondary growth thickens the stems and roots. Secondary growth results from the activity of lateral meristems. Lateral meristems are called **cambia** (cambium) and there are 2 types:

1. **Vascular cambium**—gives rise to secondary vascular tissues (secondary xylem and phloem). The vascular cambium gives rise to xylem to the inside and phloem to the outside.
2. **Cork cambium**—which forms the **periderm** (bark). The periderm replaces the epidermis in woody plants.

PLANT HORMONES

Plant growth is controlled by plant hormones, which influence cell differentiation, elongation, and division. Some plant hormones also affect the timing of reproduction and germination:

- **Auxins**—affect cell elongation (tropism), apical dominance, and fruit drop or retention. Auxins are also responsible for root development, secondary growth in the vascular cambium, inhibition of lateral branching, and fruit development. Auxin is involved in absorption of vital minerals and fall color. As a leaf reaches its maximum growth, auxin production declines. In deciduous plants this triggers a series of metabolic steps which causes the reabsorption of valuable materials (such as chlorophyll) and their transport into the branch or stem for storage during the winter months. Once chlorophyll is gone, the other pigments typical of fall color become visible.
- **Kinins**—promote cell division and tissue growth in leaf, stem, and root. Kinins are also involved in the development of chloroplasts, fruits, and flowers. In addition, they have been shown to delay senescence (aging), especially in leaves, which is one reason that florists use cytokinins on freshly cut flowers—when treated with cytokinins they remain green, protein synthesis continues, and carbohydrates do not break down.
- **Gibberellins**—produced in the root growing tips; acts as a messenger to stimulate growth, especially elongation of the stem, and can also end the dormancy period of seeds and buds by encouraging germination. Additionally, gibberellins play a role in root growth and differentiation.
- **Ethylene**—controls the ripening of fruits. Ethylene may ensure that flows are carpelate (female), while gibberellin confers maleness on flowers. It also contributes to the senescence of plants by promoting leaf loss and other changes.
- **Inhibitors**—restrain growth and maintain the period of dormancy in seeds and buds.

TROPISMS: PLANT BEHAVIOR

Tropism is the movement (and growth, in plants) of an organism in response to an external stimulus. For example, tropisms, controlled by hormones, are a unique characteristic of sessile organisms such as plants that enable them to adapt to different features of their environment—gravity, light, water, and touch—so that they can flourish. There are 3 main tropisms:

- **Phototropism**—the tendency of plants to grow or bend (move) in response to light. Phototropism results from the rapid elongation of cells on the dark side of the plant, which causes the plant to bend in the opposite direction. For example, the stems and leaves of a geranium plant growing on the windowsill always turn toward the light.
- **Gravitropism**—refers to a plant's tendency to grow toward or against gravity. A plant that displays positive gravitropism (plant roots) will grow downward, toward the center of the earth. That is, gravity causes the roots of plants to grow down so that the plant is anchored in the ground and has enough water to grow and thrive. Plants that display negative gravitropism (plant stems) will grow upward, away from the earth. Most plants are negatively gravitropic. Gravitropism is also controlled by auxin. In a horizontal root or stem, auxin is concentrated in the lower half, pulled by gravity. In a positively gravitropic plant, this auxin concentration will inhibit cell growth on the lower side, causing the stem to bend downward. In a negatively gravitropic plant, this auxin concentration will inspire cell growth on that lower side, causing the stem to bend upward.
- **Thigmotropism**—some people notice that their houseplants respond to thigmotropism (i.e., growing or bending in response to touch), growing better when they touch them and pay attention to them. Touch causes parts of the plant to thicken or coil as they touch or are touched by environmental entities. For instance, tree trunks grow thicker when exposed to strong winds and vines tend to grow straight until they encounter a substrate to wrap around.

PHOTOPERIODISM

Photoperiodism is the response of an organism (e.g., plants) to naturally occurring changes in light during a 24-hour period. Leaves are the site of photoperiod perception in plants. For instance, sunflowers are known for their photoperiodism, or their ability to open and close in response to the changing position of the sun throughout the day.

All flowering plants have been placed in 1 of 3 categories with respect to photoperiodism, short-day plants, long-day plants, and day-neutral plants:

- **Short-day plants**—flowering promoted by day lengths shorter than a certain critical day length (includes poinsettias, chrysanthemums, goldenrod, and asters)
- **Long-day plants**—flowering promoted by day lengths longer than a certain critical day length (includes spinach, lettuce, and most grains)
- **Day-neutral plants**—flowering response insensitive to day length (includes tomatoes, sunflowers, dandelions, rice, and corn)

PLANT REPRODUCTION

Plants can reproduce both sexually and asexually. Each type of reproduction has its benefits and disadvantages. A comparison of sexual and asexual plant reproduction is provided below.

Sexual Reproduction

- "Sexual reproduction occurs when a sperm nucleus from the pollen grain fuses with an egg cell from the ovary of the pistil (**pistil**—the female reproductive structures in flowers, consisting of the stigma, style, and ovary).
- "Both the sperm nucleus and the egg cell bring a complete set of genes and produce genetically unique organisms.
- "The resulting plant embryo develops inside the seed and grows when the seed is germinated" (Millen 2006).

Asexual Reproduction

- "Asexual reproduction occurs when a vegetative part of a plant, root, stem, or leaf, gives rise to a new offspring plant whose genetic content is identical to the "parent plant." An example would be a plant reproducing by root suckers—shoots that come from the root system. The breadfruit tree is an example.
- "Asexual reproduction is also called vegetative propagation. It is an important way for plant growers to get many identical plants from one very quickly.
- "By asexual reproduction, plants can spread and colonize an area quickly (e.g., crab grass)" (Millen 2006).

Animals

All animals are members of the Kingdom Animalia. With over 2 million species, Kingdom Animalia is the largest of the kingdoms in terms of its species diversity. Not surprisingly, with so much diversity among different animal species, it's difficult to imagine what they all might have in common. First, animals are composed of many cells—they are "multicellular." In most animals, these cells are organized into tissues that make up different organs and organ systems. Second, animals must get their food by eating other organisms, such as plants, fungi, and other animals—they are "heterotrophs." Third, animals are eukaryotic. Fourth, animals can be classified according to the type and presence of a coelem, an internal body cavity. In addition, all animals require oxygen for their metabolism, can sense and respond to their environment, many animals have tissues specialized for specific functions (nerve tissue, muscle), and have the capacity to reproduce sexually (though many reproduce asexually as well). During their development from a fertilized egg to adult, all animals pass through a series of embryonic stages as part of their normal life cycle.

There are 2 main types of animals, invertebrates and vertebrates. These types are discussed in the sections below.

✓   **Important Point**: The majority of the species in marine ecosystems fall under the Kingdom Animalia.

INVERTEBRATES

Invertebrates—creatures without backbones—are the most abundant creatures on Earth (more than 98% of the known animal species), crawling, flying, floating, or swimming in virtually all of Earth's habitats. Many invertebrates have a fluid-filled, hydrostatic skeleton, like the jellyfish or worm. Others have a hard outer shell, like insect crustaceans. There are many types of invertebrates. The most common invertebrates include the sponges, arachnids, insects, crustaceans, mollusks, and echinoderms.

Mollusks

U.S. Geological Survey (USGS) (1999) states that Mollusks are an amazingly diverse group of animals that live in a wide variety of environments. They can be found inhabiting trees, gardens, freshwater ponds and streams, estuaries, tidal pools, beaches, the continental shelf, and the deep ocean. Some mollusks are excellent swimmers, others crawl or burrow in mud and sand. Others remain stationary by attaching themselves to rocks, other shells, or plants, or by boring into hard surfaces, such as wood or rocks. Adult mollusks can range in size from a few millimeters (0.1 in.) to over 2 millimeters (>70 ft.) in length, as documented for some giant deep-sea squids. Their weight can vary from a few milligrams (a fraction of an ounce) to over 227 kilograms (500 lb.), as recorded for the giant South Pacific *Tridacna* clams.

The number of living species of mollusks has been estimated to range from 50,000 to 130,000. Everyone is probably familiar with some type of mollusk. They are the slugs and shelled pests in your backyard garden; the scallops, clams, mussels, or oysters on your dinner plate; the pretty shells you see washed up on the beach; the pearls or other treasures in many jewelry boxes; the octopus or squid at an aquarium (USGS 1999).

The word *Mollusca* is translated from Latin as "soft-bodied" but few physical characteristics are unique to all mollusks. The mollusks are invertebrates and therefore lack a backbone; they are unsegmented and most exhibit bilateral symmetry. Most mollusks can be described as free-living, multicellular animals that possess a true heart, and that have a calcareous exterior skeleton that covers at least the back or upper surfaces of the body. This exterior skeleton provides support for a muscular foot and the internal body organs, including the stomach mass. A thin flap of tissue called the mantle surrounds the internal organs of most mollusks, and it is this mantle that secretes the animal's shell. The nervous system of mollusks varies greatly from group to group; the clams and tusk shells have very simple nervous systems, while the squids, octopi and some other mollusks have concentrated complex nerve centers and eyes equivalent to vertebrates (USGS 1999).

 Interesting Point: Because of the many movies in which octopi and squids attack people, boats, and the like, there is a misconception that they are aggressive and dumb creatures. In fact, there are only 2 species of octopi that are aggressive (they are located in Australia), and they are highly intelligent. They are probably the most intelligent of all the invertebrates.

Annelids

Annelids are earthworms, leeches, a large number of mostly marine worms known as polychaetes (meaning "many bristles"), and other worm-like animals whose bodies are segmented. Segments each contain elements of such body systems as circulatory, nervous, and excretory tracts. Besides being segmented, the body wall of annelids is characterized by being made up of both circular and longitudinal muscle fibers surrounded by a moist, acellular cuticle that is secreted by an epidermal epithelium. All annelids except leeches also have chitonous hair-like structures, called *setae*, projecting from their cuticle. They can reproduce asexually by regeneration, but they usually reproduce sexually. There are about 9,000 species of annelids known today.

 Interesting Point: Ecologically, annelids range from passive filter feeders to voracious and active predators.

Arthropods

Insects and spiders belong to the group of animals known as **arthropods.** By nearly any measure, they are the most successful animals on the planet. They have conquered land, sea, and air, and make up over three-fourths of all currently know living organisms, or over 1 million species in all.

Arthropods have segmented bodies with jointed appendages and a chitonous exoskeleton, which must be molted and shed for growth to continue. Insect bodies are divided into 3 parts: the head, the thorax, and the abdomen. Nearly all insects have wings, and they are the only invertebrate group that can fly. Spiders and their relatives have bodies that are divided into 2 parts. The head and thorax together are called the cephalothorax, and then comes the abdomen. Most have 4 pairs of legs.

 Interesting Point: There are 200 million insects for every person on Earth.

Echinoderms

Echinoderms (from the Greek for "spiny skin") are a phylum of marine animals found at all depths. Along with spiny skin, they are characterized by an endoskeleton, radial symmetry, and a water vascular system. Echinoderms include starfish, sea stars, asteroids, sea daisies, crinoids, feather stars or seal lilies sand dollars, sea urchins, echinoids, sea cucumbers, brittle stars, and basket stars.

✓ **Interesting Point**: Echinodermata is the largest animal phylum to lack any freshwater or terrestrial representatives.

Chordata

We are most intimately familiar with the **Chordata** phylum, because it includes humans and other vertebrates. However, not all chordates are vertebrates. Chordates are defined as organisms that possess a notochord, a structure that is present during at least some part of a chordate's development. The notochord is a rod that extends most of the length of the body when it is fully developed. Other characteristics shared by chordates include the following (Hickman, Roberts, & Hickman 1994):

- Bilateral symmetry
- Segmented body, including segmented muscles
- Three germ layers and a well-developed coelom
- Single, dorsal, hollow nerve cord, usually with an enlarged anterior end (brain)
- Tail projecting beyond (posterior to) the anus at some stage of development
- Pharyngeal pouches present at some stage of development
- Ventral heart, with dorsal and ventral blood vessels and a closed blood system
- Complete digestive system
- Bony or cartilaginous endoskeleton usually present

The invertebrate chordates, which do not have a backbone, include the tunicates and lancelets. The adult form of most *tunicates* shows no resemblance to vertebrate animals, but such a resemblance is evident in the larva. The most familiar tunicates are the sea squirts. *Lancelets* are filter feeders with their tails buried in the sand and only their anterior end protruding.

VERTEBRATES

Although **vertebrates** represent only a very small percentage of all animals, their size and mobility often allow them to dominate their environments. Vertebrates include: primates, such as humans and monkeys; amphibians; reptiles; birds; and fish. Vertebrates consist of about 43,000+ species of animals with backbones. Vertebrates exhibit all of the chordate characteristics at some point during their lives. The embryonic notochord is replaced by a vertebral column in the adult. The vertebral column is made of individual hard segments (vertebrae) surrounding the dorsal hollow nerve cord. The nerve cord is the one chordate feature present in the adult phase of all vertebrates. The vertebral column, part of a flexible but strong endoskeleton, is evidence that vertebrates are segmented. The vertebrate skeleton is living tissue (either cartilage or bone) that grows as the animal grows. The post-anal tail is the only characteristic of chordates that most vertebrates keep throughout their lives.

Human Evolution

Human evolution is the biological and cultural development and change of our hominid/hominin ("hominid" term is old system; "hominin" term is new system; in this text, the new term "hominin" is used) ancestors to modern humans. Hominins evolved between 5 to 8 million years ago. To date, fossil records provide evidence of this development and date from about 4.5 million years ago. There were about 9 different hominin species. Evidence indicates that Homo sapiens made their appearance as early as 300,000 years ago.

Chapter Review Questions

1. Linnaeus's classification system is known as the _____.
2. Whittaker's 5 Kingdoms are:
3. Kingdom that comprises unicellular protozoans: _____.
4. Kingdom of unicellular molds and mildews: _____.
5. _____ are found everywhere in the environment.
6. Elongated bacteria are called _____.
7. A _____ is a virus that infects a bacterium.
8. When the host _____, phage progeny are released.
9. "First animal": _____.
10. _____ protects ciliates but still allows them to change shape: _____.
11. Life-forms involved in eutrophication of ponds and lakes are: _____.
12. The primary classification of _____ is based on cellular properties.
13. Shells of diatoms are called _____.
14. Golden-brown algae are: _____.
15. Dinoflagellates are: _____.
16. Process by which yeasts reproduce: _____.
17. Has 2 of each kind of chromosome (2n): _____.
18. Vascular plants are called _____.
19. Disk-like structures inside plant cells are called _____.
20. Lateral meristems are called _____.
21. Tendency of plants to grow or bend in response to light: _____.
22. _____ is the response of an organism to naturally occurring changes in light during a 24-hour period.
23. Creatures without backbones are: _____.
24. Earthworms and leeches are: _____.
25. Tunicate and lancelets are: _____.

THOUGHT-PROVOKING QUESTION

1. Explain biological diversity.

References and Additional Reading

Emerson, R.W. 1889. *Fortune of the Republic*. Rolling Meadows, IL: Riverside Press.

Hickman, C.P., Jr., L.S. Roberts, and F.M. Hickman. 1994. *The Biology of Animals*. St. Louis, MO: Mosby College Publishing.

Kendrick, B. 2001. *The Fifth Kingdom*, 3rd ed. Newburyport, Massachusetts: Focus Publishing.

Master Gardener. 2006. Botany: Plant Parts-Structure and Function. Accessed August 8, 2006, at http://plantfacts.ohio-state.edu/mg/manual/botany.htm.

McKinney, R.E. 1962. *Microbiology for Sanitary Engineers*. New York: McGraw-Hill.

Millen, Priscilla, 2006. *Plants in Hawaiian Environment: Botany 130*. Accessed August 26, 2006, at http://emedia.leeward.hawaii.edu/millen/bot130/learning_objectives/lo06/06.html.

NatureWorks (2007). *Angiosperms*. Durham, NH: New Hampshire Public Television.

NIH, 1994. *NIH Report on Biodiversity*. National Institutes of Health. Accessed July 2006, at www.easi.org/nape/senrep.html.

Sleigh, M.A. 1975. *Biology of Protozoa*. London: Edward Arnold.

Spellman, F.R. 2000. *Microbiology for Water/Wastewater Operators*. Boca Raton, FL: CRC Press.

Thomas, L. 1974. *The Lives of a Cell*. New York: Viking Press.

USGS. 1999. *Mollusks*. Washington, DC: U.S. Department of the Interior, U.S. Geological Survey. Accessed September 8, 2006, at http://geology.er.usgs.gov/paleo/mollusks.shtml.

Part II

HUMAN BIOLOGY

CHAPTER 7

Tissues, Organs, and Organ Systems

We are built to make mistakes, coded for error.

—Lewis Thomas (1913–1993)

Topics in This Chapter

- Tissues
- Organs and Organ Systems
- Homeostasis

Later we discuss the major human body systems. For the moment, however, we leap from the cellular level to the tissue level. In regard to the cellular level, it is important to point out that a **tissue** is primarily a group of similar cells specialized for the performance of a common function. The study of tissues is called *histology*. In humans, there are 4 main categories of tissue: epithelial tissue, connective tissue, muscle tissue, and nervous tissue. Organs represent various combinations of these 4 basic tissue types, which thus comprise the entire body. Each tissue type retains its fundamental character wherever it occurs.

Tissues

EPITHELIAL TISSUES

Epithelial tissue is a versatile tissue that covers the whole surface of the body. It is made up of cells closely packed and may act as a barrier against injury, microbial invasion, or fluid loss. This tissue is specialized to form the covering or lining of all internal and external body surfaces. The free surface of this tissue is exposed either to air or fluid. Epithelial tissue, regardless of the type, is usually separated from the underlying tissue by a thin sheet of connective tissue: **basement membrane** (a dense layer of extracellular material). The basement membrane provides structural support from the epithelium and also binds it to neighboring structures (WCSN 2000).

Epithelial tissue may be categorized by the number of layers and the shape of the free surface of the cells. Epithelial tissue which is only 1 cell thick is known as **simple epithelium**. If it is 2 or more cells thick such as the skin, it is known as **stratified epithelium**. If one layer appears to be multiple because the cells vary in length, it is known as **pseudostratified epithelium**.

The cell shapes are squamous, cuboidal, or columnar.

- **Squamous** cells have the appearance of thin, flat plates. They form the lining of cavities such as the mouth, blood vessels, heart, and lungs and make up the outer layers of the skin.
- "**Cuboidal** epithelium, as the name implies, is roughly square or cuboidal in shape. Cuboidal epithelium is found in glands and in the lining of the kidney tubules, as well as in the ducts of the glands. They also constitute the germinal epithelium which produces the egg cells in the female ovary and the sperm cells in the male testes.
- "**Columnar** epithelial cells occur in one or more layers. The cells are elongated and column-shaped. Columnar epithelium forms the lining of the stomach and intestines. Some columnar cells are specialized for sensory reception such as in the nose, ears, and the taste buds of the tongue. Unicellular glands (goblet cells) are found between the columnar epithelial cells of the duodenum. They secrete mucus or slime, a lubricating substance which keeps the surface smooth" (WCSN 2000).

Epithelial tissue performs the following functions.

- "**Protection**—Epithelial cells from the skin protect underlying tissue from mechanical injury, harmful chemicals, invading bacteria, and from excessive loss of water.
- "**Secretion**—In glands, epithelial tissue is specialized to secrete specific chemical substances such as enzymes, hormones and lubricating fluids.
- "**Sensation**—Sensory stimuli penetrate specialized epithelial cells. Specialized epithelial tissue containing sensory nerve endings is found in the skin, eyes, ears, nose, and on the tongue.
- "**Absorption**—Certain epithelial cells lining the small intestine absorb nutrients from the digestion of food.
- "**Excretion**—Epithelial tissues in the kidney excrete waste products from the body and reabsorb needed materials from the urine. Sweat is also excreted from the body by epithelial cells in the sweat glands.
- "**Diffusion**—Simple epithelium promotes the diffusion of gases, liquids, and nutrients. Because they form such a thin lining, they are ideal for the diffusion of gases (e.g., walls of lungs and capillaries).
- "**Cleaning**—Ciliated epithelium assists in removing dust particles and foreign bodies that have entered the air passages.
- "**Friction reduction**—The smooth, tightly interlocking epithelial cells that line the entire circulatory system reduce friction between the blood and the walls of the blood vessels" (WCSN 2000).

✓   **Important Point**: Dead skin cells are shed in thick sheets because they are held together by a thin layer of proteoglycan, reinforced by intermediate filaments. Such strong intercellular connections are called **desmosomes**.

CONNECTIVE TISSUES

Connective tissue functions primarily to support the body and to bind or connect together all types of tissue. Unlike epithelial tissue, connective tissue is characterized by the large amounts of intercellular substance (or matrix) that it contains. It has a sparse cell population; each cell is widely separated from the other. The various types of connective tissues are made up from 3 types of fibers.

1. "**Collagen fibers** are bundles of fiber containing 3 collagen fibers each. Collagen is the most common protein in the body. As an essential structural element in the extracellular matrix of most connective tissues, including bone and cartilage, collagen confers toughness and tensile strength. Scars are made of collagen.
2. "**Elastin** is another fibrous protein. As the name suggests, elastin is elastic—if stretched, this tissue can return to its original shape. In ordinary connective tissue, elastic fibers help restore normal shape after distortion. In high concentrations, long threads of elastin confer a yellowish color.
3. "**Reticular fibers** (from the Latin *rete*, meaning "net"), made from collagen, provide a very delicate, branched and tightly woven network, supporting individual cells in certain organs (e.g., spleen, liver)" (SIUC 2006).

✓   **Important Point**: Serous membranes reduce friction between the parietal and the visceral surfaces of an internal cavity.

Connective tissues can be classified into various types: loose connective tissue, adipose tissue, fibrous connective tissue, cartilage/bone, and blood. The standard classification scheme is based on composition—that is, on the relative proportion of various cellular and extracellular components. However, it should be pointed out that each component can vary along its own continuum:

- "**Loose connective tissue** holds organs in place and attaches the epithelium to underlying tissues. It has a relatively large proportion of ground substance, or cells, or of both cells and ground substance. In other words, loose connective tissue lacks the massive fibrous reinforcement that characterizes dense connective tissue. Nevertheless, the same types of fibers are still found, although fewer and more delicate.
- "**Adipose tissue** is loose connective tissue which is dominated by fat cells, or **adipocytes**. Since most loose connective tissue contains scattered clusters of adipocytes, the term adipose tissue is usually reserved for large masses of these cells" (SIUC). Each adipose cell stores 1 fat droplet, which varies in size.
- "**Fibrous connection tissue** contains a substantial proportion of collagen. A principal feature of fibrous tissue is flexibility combined with great tensile strength" (SIUC).

- "**Cartilage/bone** are special forms of connective tissue, made by specialized osteoblasts and chondroblasts, with uniquely solidified ground substance" (SIUC).
- **Blood** is a liquid extracellular matrix of plasma traditionally classified as a specialized form of connective tissue, with no fibers, highly fluid ground substances, and mobile cells.

Most connective tissue serves several vital functions simultaneously, including:

- Mechanical support
- Immunological defense
- Transport of nutrients and metabolites
- Instrumental in tissue repair, specifically in scar formation
- Reserve energy storage (as fat)
- Blood cell formation
- Heat generation

✓   **Important Point**: Mesolthelium is to the linings of the body cavities as endothelium is to the lungs of the heart and blood vessels.

MUSCLE TISSUE

In humans there is more **muscle tissue**, pound for pound, than any other type of tissue. Muscle tissue has an ability to relax and contract and so bring about movement and mechanical work in various parts of the body. Three types of muscles occur in humans.

- **Smooth muscle tissue** is made up of thin, elongated muscle cell fibers. The spindle-shaped, uninucleated cells contract involuntarily. Smooth muscle forms the muscle layers in the wall of hollow organs such as the digestive tract, the walls of the bladder, the uterus, various ducts of glands, and the walls of blood vessels.
- **Skeletal muscle tissue** is the most abundant tissue in the human body. These muscles are attached to and bring about the movement of the various bones of the skeleton, hence the name skeletal muscles. Skeletal muscles function in pairs to bring about the coordinated movements of the limbs, trunk, jaws, eyeballs, etc.
- **Cardiac muscle tissue** is only in the wall of the heart. Like smooth muscle tissue, it is involuntary. The cells are striated, uninucleated, and joined by intercalated disks.

NERVOUS TISSUE

Nervous tissue is specialized to react to stimuli and to conduct impulses to various organs in the body which bring about a response to the stimulus. Nerve tissue is made up of specialized nerve cells called neurons, which conduct impulses or bioelectric signals. Nervous tissue, among other important functions, allows humans to sense stimuli in both the internal and external environment.

Organs and Organ Systems

Tissues work together to perform a particular task. An **organ** (Latin: *organum*, meaning "tool, instrument") is a group of tissues that perform a specific task or group of tasks. Human organs include the heart, lungs, brain, eye, stomach, spleen, pancreas, kidneys, liver, intestines, skin, uterus, and bladder. Human organs inside the thorax or abdomen are often referred to as internal organs, sometimes called "the viscera."

An **organ system** is a group of related organs. Organs within a system may be related in any number of ways, but relationships of function are most common. For example, the stomach is part of the digestive system. The digestive system is a continuous tube beginning with the mouth and ending at the rectum. In between are several other components or organs. Another example is the urinary system, which is composed of organs that work together to produce, store, and carry urine.

BIOENERGETICS

Organ systems do not run on perpetual motion; instead, they require energy that is balanced in terms of intake and loss. **Bioenergetics** is the study of energy investment and flow through living systems. It includes the study of thousands of different processes ranging from cellular respiration and the production of ATP. Humans get the fuel needed for the production of energy by eating.

Homeostasis

Simply, **homeostasis** is the ability of an organism (including humans) to maintain the internal environment within tolerable limits—the narrow range of conditions where cellular processes are able to function at a level consistent with the continuation of life. This internal environment must be maintained in response to changes in the conditions of both the external environment and internal environment.

Feedback mechanisms control and adjust chemical and physical aspects of the body within tolerable ranges. That is, feedback mechanisms are the general mechanism of nervous or hormonal regulation in humans. Basically, feedback occurs when the response to a stimulus has an effect of some kind on the original stimulus. The nature of the response determines how the feedback is labeled:

- **Negative feedback** is when the response diminishes the original stimulus. Negative feedback is most common in biological systems. For example, physical work creates metabolic heat which raises the body temperature (the stimulus), cooling mechanisms such as vasodilation, or flushed skin, and sweating begin (the response), and body temperature falls (thus decreasing the original stimulus).
- **Positive feedback** is when the response enhances the original stimulus. Positive feedback is less common, understandably, as most changes to steady state pose a threat, and to enhance them would be most unhelpful.

Chapter Review Questions

1. Name the 4 primary tissue types.
2. Which primary tissue type always has a free surface exposed either to the internal or external environment?
3. Four functions of the epithelium include:
4. The dominant fiber type in dense connective tissue is _____.
5. Osteoblasts are associated with _____.
6. The study of tissues is called _____.
7. Smooth tissue is characteristic of some _____ tissue.
8. Cells that store fat are called _____.
9. The reduction of friction between the parietal and the visceral surfaces of an internal cavity is the function of _____ membranes.

References and Additional Reading

Cheraskin, E. 1999. *Human Health and Homeostasis*. New York: Clayton College/Natural Reader Press.

Clancy, J. 2002. *Human Biology*, 4th ed. New York: Jones and Bartlett Publishers, Inc.

Hacker, G.W. 2004. *Molecular Morphology of Human Tissue with Light Microscopy*. Boca Raton, FL: CRC Press.

SIUC 2006. *Connective Tissue Study Guide*. Accessed August 8, 2006, at www.siumed.edu/~dking2.

Thomas, L. 1976. *The Medusa and the Snail (To Err Is Human)*. New York: Penguin.

WCSN. 2000. *Epithelial Tissue*. Cape Town, South Africa: Western Cape Schools Network: Internet Bio-ed Project. Accessed August 10, 2006, at www.botany.uwc.ac.za/sc_ed/.

CHAPTER 8

Integumentary System

Each square inch of human skin consists of 20 feet of blood vessels.

Topics in This Chapter

- Skin
- Skin Functions
- Skin Color
- Skin Appendages
- Nerve Endings in the Skin
- Homeostatic Imbalances of the Skin

All body systems work in an interconnected manner to maintain the internal conditions (homeostasis) essential to the function of the body. The integumentary system has multiple roles in maintaining homeostasis, including protection temperature regulation, sensory reception, biochemical synthesis, and absorption. The skin (also known as the **cutis** or **integument**) serves as the main cover of the body and thus plays a major role in maintaining homeostasis. The skin is considered the largest organ in the body: 12%–16% of total body weight, with a surface area of 1 to 2 meters. Of all the body's organs, none is more easily exposed to infection, disease, and injury than the skin.

Skin

The skin consists of 2 main layers. The part you see and touch is the epidermis—the outer, thinner portion composed of layers of epithelium. The dermis is the inner, thicker layer of connective tissue. The dermis is attached to fibers of a layer called the hypodermis (subcutaneous layer), which isn't considered part of the skin. The subcutaneous layer serves largely as a storage depot for fat (adipose tissue) and contains large blood vessels that supply the skin. The subcutaneous layer is attached to the underlying tissues.

EPIDERMIS

The **epidermis** is the outermost layer of the skin and consists of stratified squamous epithelium with a layer of keratin. Four cell types, with different embryologic origins, are distinguished in the epidermis:

- Keratinocytes (keratin production)
- Melanocytes (pigment production)
- Langerhans cells (immune system)
- Merkel cells (diffuse neuroendocrine system)

Keratinocytes are the dominant cell type of the epidermis (about 90%), and are part of the keratinizing system, involved in the cornification or keratinization of the skin. The most superficial cells are essentially dead cells, or scales, primarily composed of keratin. These superficial keratinized cells are continuously being lost and need to be replaced. New keratinocytes are continuously formed in the basal layers of the epithelium.

Roughly 8% of the epidermal cells are **melanocytes,** which absorb ultraviolet light and produce the main component of skin coloration, **melanin**. Melanocytes are found in the more basal layers of the epidermis. They are rounded cells with processes that extend between the adjacent keratinocytes.

Langerhans cells are found in the epidermis. They are star-shaped cells found mainly in the spiny layer. They function as antigen-presenting cells of the immune system of the skin. That is, they participate in immune responses mounted against microorganisms.

Merkel cells are least numerous of the epidermal cells and are located in the deepest layer of the epidermis where they are in contact with nerve cells and function in the sensation of touch.

The epidermis of thick skin, depending on location (e.g., palms of hands, soles of feet, etc.), is 5 layers thick:

- **Basal layer** (*Stratum basalis*)—the basal layer is the innermost layer of the epidermis, and contains small round cells called basal cells. The epithelial cells of the basal layer are closest to the basal lamina and to the underlying dermis. Keratinocyte and melanocyte cells are contained in the basal layer. These cells are typically columnar or cuboidal with many examples of mitosis.
- **Spiny layer** (*Stratum spinosum*)—the keratinocytes of the spiny layer are more rounded or oval and at the light microscope level there appear to be pronounced gaps between the cells. These spines are composed of prominent desmosomes.
- **Granular layer** (*Stratum granulosum*)—this layer consists of 2–5 rows of fairly flattened cells characterized by large cytoplasmic basophilic granules known as **keratohyalin granules**.
- **Clear layer** (*Stratum lucidum*)—this is seen as a thin, undulating layer, strained well with eosin.

- **Horny layer** (*Stratum corneum*)—the stratum corneum is the outermost layer of the epidermis, and is made up of 10 to 30 thin layers of continually shedding, dead keratinocytes. The cells of the horny layer (so named because its cells are toughened like an animal's horn) are flattened and their nuclei can no longer be distinguished. The cells show a thickening of their plasmalemma. The cells are packed with **keratin**.

✓   **Important Point**: The basal layer and the spiny layer together are known as the **Malphighian layer** or **germinative layer**.

DERMIS

The **dermis** is the layer of connective tissue beneath and is the thickest of the three layers of the skin, making up approximately 9% of the thickness of the skin. The epidermis is derived from mesoderm. The main functions of the dermis are to regulate temperature and to supply the epidermis with nutrient-saturated blood. Much of the body's water supply is stored within the dermis. This layer contains most of the skins' specialized cells and structures, including blood vessels, lymph vessels, hair follicles, sweat glands, sebaceous glands, nerve endings, collagen and elastin.

The dermis layer is made up of 2 sublayers:

- **papillary layer**—contains a thin arrangement of collagen fibers. The papillary layer supplies nutrients to select layers of the epidermis and regulates temperature. Both of these functions are accomplished with a thin, extensive vascular system that operates similarly to other vascular systems in the body. Constriction and expansion control the amount of blood that flows through the skin and dictate whether body heat is dispelled when the skin is hot or conserved with it is cold.
- **reticular layer**—is thicker than the papillary layer and consists of a denser, irregular connective tissue composed mainly of collagen bundles that strengthens the skin, providing structure and elasticity. It also supports other components of the skin, such as hair follicles, sweat glands, and sebaceous glands. The term *reticular* refers to the netlike arrangement of these bundles.

✓   **Important Point**: Within the reticular layer of the dermis are found fibroblasts, blood vessels, macrophages, and sensory receptors.

HYPODERMIS

"The hypodermis is the deepest layer of the skin. It is manufactured by specialist cells and is composed mainly of fat (adipose tissue). The thickness of this layer varies from person to person and also from one body area to the next, with very little around the spine and nose, but with more where curves are formed. The hypodermis in women is thicker than in men, which helps to form the rounded curves in women. This layer acts as insulation, protects the internal organs from temperature variations, and also acts as an energy reserve from which the body can draw as required" (Dermaxime 2006).

Skin Functions

The human skin functions as follows:

- Provides **mechanical protection** against external abrasions or injury and against invasion of foreign objects
- Provides **thermoregulation** of the body
- Provides **osmoregulation** of body fluids and ions
- Provides exocrine glands for **excretion** and **secretion**
- Provides **sensory reception** from the external environment
- Provides **metabolic functions**
- Provides route of **absorption**

Skin Color

Human skin color is quite variable around the world. It ranges from a very dark brown among some Africans, Australians, and Melanesians to a near yellow-pink among some northern Europeans. There are no people who actually have true black, white, red, or yellow skin. These are commonly used color terms that do not reflect reality.

Coleman (2006) points out that the "color of the skin depends on its thickness, and the degree of underlying vascularization, especially the oxyhemoglobin. If blood vessels are contracted, as in cold environments, the skin is more pallid, whereas in hot conditions or if we exert ourselves the vessels dilate and we appear redder. The yellowish color of skin is due to the pigment carotene. Skin color can sometimes provide diagnostic clues to underlying disorders (anemia, cyanosis, hepatitis)."

Skin color is due primarily to the presence of a pigment called **melanin**. Both light and dark complexioned people have this pigment. However, 2 forms are produced, one of which is red to yellow in color, and the other dark brown to black.

Melanin is normally located in the epidermis, or outer skin layer. It is produced at the base of the epidermis by specialized cells called **melanocytes**. Melanin is lacking in thick skin. The palms of the hands and the soles of the feet are unpigmented, even in dark-skinned people.

Skin Appendages

Skin appendages (derived from epithelium) include:

- Hair
- Nails
- Exocrine glands (sebaceous glands, sweat glands)

Hair is one of the unique characteristics of mammals. Hairs are thin filaments of keratin, modified epidermal cells that develop in the dermis from epithelial invaginations of the epidermis, which interact with a germinative center of the dermis—the dermal papilla. The hair shaft extends above the skin surface; the hair root extends from the surface to the base of the hair bulb. Genetics control several features of hair: baldness, color, texture. Hair grows discontinuously, with periods of growth followed by periods of inactivity.

Nails consist of horny plates of highly keratinized, modified epidermal cells. The nail arises from the nail bed, which is thickened to form a lunula (commonly called the "little moon"). Cells forming the nail bed are linked together to form the nail.

"**Sebaceous glands** (not present in thick skin) are dermal exocrine glands associated with hairs which secrete an oily substance (**sebum**) on the growing hair. The sebum is important in maintaining the flexibility of the hair. The glands are composed of alveoli and a short secretory duct. The outermost cells of the gland are filled with lipid droplets. These cells are secreted in their entirety (**holocrine secretion**). The extrusion of the sebum results from the contraction of the arrector pili muscles. The sebum secretion is influenced by sex hormones (**androgens** and **estrogens**) and, during the hormonal imbalances of puberty, these secretions may result in adolescent acne" (Coleman 2006).

Sweat glands are "simple exocrine glands that secrete sweat. The **secretory units** are simple convoluted tubular epithelial structures in the dermis and a straight **secretory duct**. The convolutions of the secretory unit are seen in histological preparations as many associated profiles of the same unit. The secretory ducts are surrounded by myoepithelial cells, which on contraction cause the expulsion of the sweat" (Coleman 2006).

There are 2 types of sweat glands:

- **Eccrine sweat glands** (merocrine sweat glands)—produce sweat—a mixture of water and salts. Sweat plays an important part in regulating the temperature of the body by cooling it by evaporation of water from the skin. It also provides a useful natural method of removing waste products (toxins) from the body. The tiny ducts of eccrine glands pass through the dermis and epidermis and empty directly onto the skin.
- **Apocrine sweat glands**—are "large sweat glands located in the axilla (armpit) and also in association with the eternal genitalia and anus. The apocrine sweat glands only become functional at puberty and are influenced by sex steroids" (Coleman 2006).

Nerve Endings in the Skin

The skin has many sensory elements that respond to external impulses and signals.

- **Free nerve endings**—are nonencapsulated, unspecialized, afferent nerve endings, meaning they bring information, such as indication of pressure or something painful, from the body's periphery to the brain. They are the most common type of nerve ending, and are most frequently found in the skin.

- **Meissner corpuscles**—are mechanoreceptors (tactile corpuscles) present in the dermal papilla. They are distributed throughout the skin, but concentrated in areas especially sensitive to light touch, such as the fingertips, palms, soles, lips, tongue, face, nipples, and the external skin of the male and female genitals.
- **Merkel corpuscles**—are mechanoreceptors located in the epidermis surrounding hair follicles responsive to touch.
- **Pacinian corpuscles**—are ovoid-shaped, found in the dermis of thick skin of fingers, and respond to pressure and vibration.
- **Krause end bulbs**—are found in the dermis and respond to cold.

Homeostatic Imbalances of the Skin

The most common skin disorders result from bacterial, viral, or fungal infections or allergies. For example, **athlete's foot** results from a fungal infection. **Contact dermatitis** is caused by exposure of the skin to chemicals that provoke allergic responses in sensitive individuals. **Acne** is an inflammation of oil glands that usually begins at puberty when the oil glands grow in size and increase the production of sebum. Bacteria can colonize sebaceous follicles and cause infections. **Psoriasis** is a chronic, non-contagious relapsing skin condition characterized by red skin elevations covered with scales. The most common cause of **skin cancer** is exposure to the sun (UV radiation). There are several types of cancer that start in the skin. The most common are basal cell cancer and squamous cell cancer. **Melanoma** is a disease (malignancy) of the skin in which cancer (malignant) cells are found in the cells that color the skin (melanocytes). Melanoma occurs most frequently in white people, and is rare in people with dark skin; it is usually found in adults, though occasionally melanoma may develop in children and adolescents. Melanoma is a more serious type of cancer than the more common skin cancers, basal cell cancer or squamous cell cancer. **Basal cell carcinoma** is where the basal cells become cancerous. This is the most common type of skin cancer and is usually highly curable when detected early. **Squamous cell carcinoma** is a type of skin cancer arising in squamous cells. Cure rates are very high when detected and treated early.

Chapter Review Questions

1. Name the 4 major components of the integumentary system.
2. The skin conserves heat by reducing its secretion of sweat and constriction of its _____ vessels.
3. Name the layers of the dermis.
4. In skin layers, the protein _____ forms a barrier to water-soluble substances.
5. Melanin protects the _____ from sun damage.
6. Where are the Langerhans cells located within the skin?

7. List 6 functions of the integumentary system.
8. What accessory structure are arrector pili muscles associated with?
9. What 2 types of cells are found in the stratum basalis?
10. Name the 3 factors that contribute to skin color.

References and Additional Reading

Coleman, R. 2006. *The Skin*. Haifa, Israel: Technion-Israel Institute of Technology. Accessed August 11, 2006, at www.technion.ac.il/~mdcourse/274203/lect12.html.

Dermaxime. 2006. The Skin—Its Configuration and Function. Accessed August 8, 2006, at www.dermaxime.com/skin.htm.

Leonardi, P. 2005. *Anatomy and Physiology Study Guide, Vol. 1*. Silver Educational Publishing.

Marieb, E.N. 2003. *Human Anatomy & Physiology,* 6th ed. Benjamin Cummings.

Martini, F.H. 2005. *Fundamentals of Anatomy & Physiology*, 7th ed. Benjamin Cummings.

Seeley, R.R., et al. 2004. *Anatomy and Physiology*. New York: McGraw-Hill.

Seeley, R.R., et al. 2006. *Essentials of Anatomy and Physiology*. New York: McGraw-Hill.

CHAPTER 9

Circulatory System

I think it's liquid aggravation that circulates through his veins, and not regular blood.

—Charles Dickens (1812–1870)

Topics in This Chapter

- Blood
- Blood Vessels
- Heart
- Circulatory System
- Diseases of the Circulatory System

As humans evolved to the complex organisms we know today, their central body plans required a circulatory system that could transport materials such as nutrients, oxygen, and waste products throughout the body. The closed circulatory system of humans is composed of 3 components: blood, blood vessels, and the heart. In this chapter, we review each of these vital circulatory system components.

Blood

Blood is the only liquid connective tissue in the body. "The average adult has about 5 liters of blood inside of their body, coursing through vessels, delivering essential elements, and removing harmful wastes. Blood, the fluid of life, transports oxygen from the lungs to body tissue and transports carbon dioxide from body tissue to the lungs. Blood is the fluid of growth, transporting nourishment from digestion and hormones from glands throughout the body. Blood is the fluid of health, transporting disease-fighting substances to the tissue and waste to the kidneys. Red blood cells (RBCs) and white blood cells (WBCs) are responsible for nourishing and cleansing the body. Since the blood cells are alive (i.e., they contain living cells), they too need nourishment. Vitamins and minerals keep the blood healthy. The blood cells have a definite life cycle, just as all living organisms do" (Resources 2006).

PLASMA

Approximately 55% of blood is **plasma**, a straw-colored clear liquid, which is mostly water. The plasma transports nutrients, waste products of metabolism, respiratory gases, and hormones. The liquid plasma carries the solid cells and the *platelets* which help blood clot. Without blood platelets, we would bleed to death. *Blood proteins*, also called serum proteins, are proteins found in blood plasma. There are 3 main plasma proteins:

- **Albumin**—is the smallest and most numerous of the plasma proteins. It transports some of the steroid hormones and helps recover water that has been lost through the capillaries.
- **Immunoglobulins** (antibodies)—are glycoproteins in the immunoglobulin family that function as antibodies; they help in the immune system by attacking bacteria and viruses. They are found in the blood and tissue fluids, as well as many secretions. They are synthesized and secreted by plasma cells that are derived from the B cells of the immune system.
- **Fibrinogen**—is a soluble protein that circulates in the blood and provides the material from which the insoluble fibrin clot is formed during blood coagulation. They play key roles in osmotic balance, pH buffering, and the regulation of membrane permeability. High plasma fibrinogen concentration predicts future coronary heart disease in men and women. Fibrinogen, as an acute phase reactant, responds to infection and other short-term inflammatory stressors such as tobacco smoke.

BLOOD CELLS

As mentioned, "the blood consists of a suspension of special cells called plasma. Blood consists of 55% plasma, and 45% formed cells (or formed elements). Special cells in the blood are classified as erythrocytes (RBCs), leukocytes (WBCs), and platelets (small fragments of WBCs)" (Fun Science 2006).

"**Erythrocytes**, or **RBCs**, are the most abundant blood cells. In humans, RBCs are devoid of a nucleus and have the shape of a biconcave lens. RBCs are rich in hemoglobin, a protein able to bind in a faint manner to oxygen. Hence, these cells are responsible for providing oxygen to tissues and partly or recovering carbon dioxide produced as waste" (Fun Science 2006).

✓   **Important Point:** The life span of an RBC is only 120 days, after which they are destroyed in the liver and spleen.

Leukocytes, or **WBCs**, are responsible for the defense of the organism. In the blood, they are much less numerous than RBCs. Leukocytes divide in 2 categories: granulocytes and lymphoid cells. The term *granulocyte* comes from the presence of granules in the cytoplasm of these cells. In the different types of granulocytes, the granules are different and help us to distinguish them. There are 5 different types of leukocytes, in different proportions, present in the blood:

- **Neutrophils** (50%–70%) are very active in phagocyting bacteria and are present in large amount in the pus of wounds.
- **Eosinophils** (2%–4%) attack parasites and phagocyte antigen-antibody complexes.
- **Basophils** (0.5%–1%) secrete anticoagulant and vasodilatory substances as histamines and serotonin.
- **Lymphocytes** (20%–40%) are the main constituents of the immune system, which is a defense against the attack of pathogenic microorganisms such as viruses, bacteria, fungi, and protista.
- **Monocytes** (3%–8%) are the precursors of macrophages.

✓   **Important Point**: WBCs are made from stem cells in bone marrow.

Platelets are irregularly shaped, colorless bodies that are present in blood. They are not really cells; they are packets of cytoplasm (cell fragments) that release the enzyme thromboplastin when they come into contact with a foreign substance within the blood or the rough edges of an open wound. Their sticky surface lets them, along with other substances, form clots to stop bleeding. When bleeding from a wound suddenly occurs, the platelets gather at the wound and attempt to block the blood flow. The mineral calcium, vitamin K, and the protein fibrinogen (converted to fibrin) help the platelets form a clot.

✓   **Important Point**: Platelets survive for 10 days before being removed by the liver and spleen.

BLOOD CLOTTING FACTORS

When an injury occurs that causes bleeding, a blood clot (a gel that contains formed elements of the blood) will develop to stop the bleeding. When the injured blood vessel has healed itself, the clot is no longer needed and it is reabsorbed by the body. Proteins in the blood called **clotting factors** work to ensure that blood does not become too thick or thin. Clotting is a complex cascade of events in which one of several clotting factors activates the next one in a fixed sequence. Under certain circumstances the body becomes more prone to developing blood clots. This can lead to blood clots that can cause harm in the heart, lungs, brain, or extremities. If a clot forms in such an area, it might cause a blockage in the small blood vessels and hinder blood flow. More seriously, a piece of the clot can break off and travel through the circulatory system and get stuck in the smaller blood vessels of the lungs, brain, or heart. This can be a life-threatening event.

Blood Vessels

Blood vessels are the channels or conduits through which blood is distributed to body tissues. The vessels make up 2 closed systems of tubes that begin and end at the heart. One system, the pulmonary system, transports blood from the right ventricle to the lungs and back to the left atrium. The other system, the systemic vessels, carries blood from the left ventricle to the tissues in all parts of the body and then returns the blood to the right atrium. Blood vessels are classified as arteries, capillaries, or veins.

ARTERIES

Arteries carry blood away from the heart and have thick, complex, elastic (able to expand and contract), muscular walls that can dilate or contract to control blood pressure within the vessels. Pulmonary arteries transport blood that has low oxygen content from the right ventricle to the lungs. Systemic arteries transport oxygenated blood from the left ventricle to the body tissues. Blood is pumped from the ventricles into large elastic arteries that branch repeatedly into smaller and smaller arteries until the branching results in microscopic arteries called **arterioles**. The arterioles play a key role in regulating blood flow into the tissue capillaries. About 10% of the total blood volume is in the systemic arterial system at any given time.

Arteries have 3 layers of thick walls. The innermost layer, the *tunica intima*, is simple squamous epithelium surrounded by a connective tissue basement membrane with elastic fibers. The middle layer, the *tunica media*, is primarily smooth muscle and is usually the thickest layer. It not only provides support for the vessel but also changes vessel diameter to regulated blood flow and blood pressure. The outermost layer, which attaches the vessel to the surrounding tissue, is the *tunica externa*. This layer is connective tissue with varying amounts of elastic and collagenous fibers. The connective tissue in this layer is quite dense where it is adjacent to the tunica media, but it changes to loose connective near the periphery of the vessel.

CAPILLARIES

Arteries are too large to service every little cell in the body. As arteries get farther from the heart, they begin to branch into smaller and smaller vessels, which eventually branch into thousands of capillaries. **Capillaries** are the smallest (as small as one cell thick) and most numerous of the blood vessels; they form the connection between the vessels that carry blood away from the heart (arteries) and the vessels that return blood to the heart (veins). The primary function of capillaries is the exchange of materials between the blood and tissue cells.

Capillary distribution varies with the metabolic activity of body tissues. Tissues such as skeletal muscle, liver, and kidney have extensive capillary networks because they are metabolically active and require an abundant supply of oxygen and nutrients. Other tissues, such as connective tissue, have a less abundant supply of capillaries. The epidermis of the skin and the lens and cornea of the eye completely lack a capillary network. About 5% of the total blood volume is in the system capillaries at any given time, another 10% is in the lungs. After providing nutrients and oxygen and picking up waste, capillaries begin to merge into larger and larger vessels called **venules**, eventually converging into veins.

✓   **Important Point**: Smooth muscle cells in the arterioles where they branch to form capillaries regulate blood flow from the arterioles into the capillaries.

VEINS

Veins carry blood toward the heart. After blood passes through the capillaries, it enters the smallest veins, called *venules*. From the venules, it flows into progressively larger and larger veins until it reaches the heart. The blood in the veins is not pushed by pumping of the heart, so the blood pressure and forward momentum of the blood in veins is lower than in arteries. Blood in veins is largely pushed along by the contractions of the skeletal muscles as the body moves around. To ensure that the blood in veins flows toward the heart, veins contain unidirectional valves (the valves prevent back-flow of blood). In the pulmonary circuit, the pulmonary veins transport blood from the lungs to the left atrium of the heart. This blood has high oxygen content because it has just been oxygenated in the lungs. Systemic veins transport blood from the body tissue to the right atrium of the heart. This blood has reduced oxygen content because the oxygen has been used for metabolic activities in the tissue cells.

The walls of the veins have the same 3 layers as the arteries. Although all the layers are present, there is less smooth muscle and connective tissue. This makes the walls of veins thinner than those of arteries, which, again, is related to the fact that blood in the veins has less pressure than in the arteries. Because the walls of the veins are thinner and less rigid than arteries, veins can hold more blood. Almost 70% of the total blood volume is in the veins at any given time.

Heart

The human heart is a 2-sided, 4-chambered, shell-like structure with muscular walls. Two chambers are called the atria. The other 2 are ventricles. The 2 atria form the curved top of the heart. The ventricles meet at the bottom of the heart to form a pointed base, which points toward the left side of the chest. The left ventricle contracts most forcefully. The top of the heart connects to a few large blood vessels. The largest of these is the aorta, or main artery, which carries nutrient-rich blood away from the heart. Another important vessel is the pulmonary artery, which connects the heart with the lungs as part of the pulmonary circulation system. The 2 largest veins that carry blood into the heart are the superior vena cava and the inferior vena cava (heart's veins). The superior is located near the top of the heart. The inferior is located beneath the superior.

The heart's structure makes it an efficient pump. The average heart's muscle, called cardiac muscle, contracts and relaxes about 70 times per minute. As the cardiac muscle contracts it pushes blood through the chambers and into the vessels. Nerves connected to the heart regulate the speed with which the muscle contracts. Heartbeat is controlled by the autonomic nervous system.

✓   **Interesting Point**: Over a lifetime of 70 years, the heart beats approximately 2.5 billion times and pumps 200 million liters of blood.

The heart, considering how much work it does, is surprisingly small. The average adult heart is about the size of a clenched fist and weighs about 11 ounces. The heart is well protected. It is located in the middle of the chest behind the breastbone, between the lungs. The heart rests in a moistened chamber called the pericardial cavity, which is surrounded by the ribcage. The diaphragm, a tough layer of muscle, lies below.

Circulatory System

With each beat, the heart pumps blood into 2 closed circuits, the systemic circuit and the pulmonary circuit. A third type of circulation is known as the portal system—specialized channels that connect one capillary bed site to another but do not depend directly on a central pump. The largest of these is the hepatic portal system, which connects the intestines to the liver.

- The **systemic circulation system** transfers oxygenated blood from the left side of the heart via the aorta to all of the body tissues and returns deoxygenated blood with high carbon dioxide content from the tissues to the heart.
- The **pulmonary circulation system** pumps blood from the right side of the heart through a single artery, the pulmonary artery, which divides into 2—1 branch supplying each lung. It receives all the deoxygenated blood returning from the systemic circulation.
- In the **portal circulation system**, there are 2 capillary beds, one after the other, in series. Portal circulation refers to the circulation of blood from the small intestine to the liver, via the portal vein.

Diseases of the Circulatory System

We all know how important the circulatory system is to us in maintaining a healthy lifestyle and to life itself. Accordingly, it is no wonder that we worry when we hear someone has circulatory system problems. As mentioned, the heart is the center of the circulatory system. Through the body's blood vessels, the heart pumps blood to all of the body's cells. The blood carries oxygen, which the cells need. Circulatory disease (cardiovascular disease) is a group of problems that occur when the heart and blood vessels are not working the way they should.

Some of the problems that go along with circulatory disease are listed and described below:

- **Hypertension**—the human heart beats about 100,000 times a day. Each time the heart beats, blood is pumped through the various blood vessels in the body. Blood pressure is defined as the pressure the blood exerts on the vessel walls. If the pressure is too low, the blood may not reach certain areas of the body. If the blood pressure is too high, the heart and blood vessel walls may be damaged. *Hypertension* is a condition where the blood pressure is constantly higher than normal. This poses a serious health risk because it forces the heart to work extra hard. The effects of hypertension include strokes and heart attacks. Approximately 25% of Americans suffer from hypertension. This disturbing statistic is related in large part to contributing factors such as smoking; a diet rich in salt, fat and cholesterol; and stress.

  ✓   **Important Point**: In 95% of hypertension cases, the cause is unknown. Experts are aware of contributing factors that increase the chances of developing hypertension, but no one has been able to determine a direct link between the factors and developing high blood pressure.

- **Arteriosclerosis/Atherosclerosis**—*arteriosclerosis*, also called hardening of the arteries, means a person's arteries become thickened and are no longer as flexible. People with *atherosclerosis* have a buildup of cholesterol and fat (plaque) that makes their arteries narrower so less blood can flow through. It usually affects large- and medium-sized arteries. Plaques that rupture cause blood clots to form that can block blood flow or break off and travel to another part of the body. If either happens and blocks a blood vessel that feeds the heart, it causes a heart attack. If it blocks a blood vessel that feeds the brain, it causes a stroke. Some hardening of arteries often occurs when people grow older.

- **Heart attack**—occurs when the supply of blood and oxygen to an area of heart muscle is blocked, usually by a clot in a coronary artery. Often, this blockage leads to arrhythmias (irregular heartbeat or rhythm) that cause a severe decrease in the pumping function of the heart and may bring about sudden death. If the blockage is not treated within a few hours, the affected heart muscle will die and be replaced by scar tissue. A heart attack is a life-threatening event. Everyone should know the warning signs of a heart attack and how to get emergency help. Many people suffer permanent damage to their hearts or die because they do not get help immediately. Each year, more than a million people in the U.S. have a heart attack and about half (515,000) of them die.

  ✓   **Important Point**: About half of those who suffer a heart attack die within 1 hour of the start of symptoms and before reaching the hospital.

- **Leukemia**—is a malignant disease (cancer) of the bone marrow and blood, characterized by an abnormal proliferation of blood cells, usually WBCs.

Chapter Review Questions

1. The only liquid connective tissue in the body: _____.
2. Makes up 55% of blood: _____.
3. Smallest and most plentiful blood proteins: _____.
4. _____ provide oxygen to tissues and recover carbon dioxide.
5. _____ are responsible for the defense of the organism.
6. Precursors of macrophages: _____.
7. It is a gel containing formed elements of the blood: _____.
8. _____ carry blood away from the heart.
9. Microscopic arteries: _____.
10. The 2 largest veins of the heart: _____.

References and Additional Reading

American Heart Association, 2002. *2002 Heart and Stroke Statistical Update*. Dallas, TX: American Heart Association.

Bullock, B.A., and L.H. Reet. 2000. *Focus on Pathophysiology*. Baltimore: Lippincott.

Corwin, E.J. 2000. *Handbook of Pathophysiology*. Baltimore: Lippincott.

Dickens, C., 1995. *Martin Chuzzlewit*. New York: Everyman's Library.

Fun Science Gallery. 2006. *Blood Cells*. Accessible at www.funsci.com/fun3_en/blood.

Hansen, M. 1998. *Pathophysiology: Foundations of Disease and Clinical Intervention*. Philadelphia: Saunders.

Hartshorn, J.C., M.L. Sole, and M.L. Lamborn. 1997. *Introduction to Critical Care Nursing*. Philadelphia: Saunders.

Huether, S.E., and K.L. McCance. 2002. *Pathophysiology*. St Louis, MO: Mosby.

Resources for Science Planning. 2006. *Lifeblood*. Accessible at www.fi.edu/bioscie/blood.

SEER. 2000. *Blood*. Atlanta, GA: U.S. National Cancer Institute's Surveillance, Epidemiology and End Results (SEER) program. Accessed August 12, 2006, at www.training.seer.cancer.gov.

Sparknotes. 2006. *The Circulatory System*. Accessed August 12, 2006, at www.sparknotes.com.

CHAPTER 10

Lymphatic System

The lymphatic system becomes a crucial battleground during infection.

Topics in This Chapter

- The Immune System
- Structure of the Immune System
- Immune Cells
- Immune Response
- Natural and Acquired Immunity
- Disorders of the Immune System

Scientists have learned much about the immune system, and continue to study how the body launches attacks that destroy invading microbes, infected cells, and tumors while ignoring healthy tissues. New technologies for identifying individual immune cells are now letting scientists quickly determine which targets are triggering an immune response. Improvements in microscopy are permitting the first-ever observations of B cells, T cells, and other cells as they interact within lymph nodes and other body tissues. Moreover, scientists are rapidly unraveling the genetic blueprints that direct the human immune response as well as those that dictate the biology of bacteria, viruses, and parasites (NIH 2003).

In regard to exposure to organisms (bacteria, parasites, and viruses), there is no way we can avoid them. Each day, we are literally exposed to thousands of pathogens—disease-causing organisms. Fortunately for us, our bodies are equipped with a variety of defenses against pathogens.

Note: Much of the material in this chapter is from or is based on NIH's 2003 *Understanding the Immune System: How It Works* and the SEER anatomy and physiology training module *Functions of the Lymphatic System*.

Our primary defense mechanism against pathogens is the **lymphatic system**. The lymphatic system is the body's second circulatory system—a system of vessels transporting **lymph** throughout the body. These vessels consist of organs, ducts, and nodes. The lymphatic system transports lymph (a watery clear fluid). This fluid distributes immune cells and other factors throughout the body. It also interacts with the blood circulatory system to drain fluid from cells and tissues. The lymphatic system contains immune cells called lymphocytes, which protect the body against antigens that invade the body.

The Immune System

The **immune system** is a network of cells, tissue, and organs that work together to defend the body against attacks by "foreign" invaders. These are primarily microbes (germs)—tiny, infection-causing organisms such as bacteria, viruses, parasites and fungi. Because the human body provides an ideal environment for many microbes, they try to break in. It is the immune system's job to keep them out or, failing that, to seek out and destroy them. However, when the immune system hits the wrong target or is crippled, it can unleash a torrent of diseases, including allergy, arthritis, or AIDS.

The immune system is amazingly complex. It can recognize and remember millions of different enemies, and it can produce secretions and cells to match up with and wipe out each one of them.

The secret to its success is an elaborate and dynamic communications network. Millions and millions of cells, organized into sets and subsets, gather like clouds of bees swarming around a hive and pass information back and forth. Once immune cells receive the alarm, they undergo tactical changes and begin to produce powerful chemicals. These substances allow the cells to regulate their own growth and behavior, enlist their fellows, and direct new recruits to trouble spots.

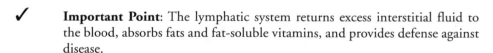

✓   **Important Point**: The lymphatic system returns excess interstitial fluid to the blood, absorbs fats and fat-soluble vitamins, and provides defense against disease.

DEFINITION OF KEY TERMS (NIH 2003)

Before proceeding with an in-depth discussion of the immune system, for better understanding, it is important to list and define pertinent terms associated with the immune system.

- **AIDS** (acquired immunodeficiency syndrome)—is a collection of disorders resulting from the destruction of T cells by the Human Immunodeficiency Virus (HIV), a retrovirus.
- **Adenoids**—see tonsils.

- **Adrenal gland**—a gland located on each kidney that secretes hormones regulating metabolism, sexual function, water balance, and stress.
- **Allergen**—any substance that causes an allergy.
- **Allergy**—a harmful response of the immune system to normally harmless substances.
- **Antibodies**—molecules (also called immunoglobulins) produced by a B cell in response to an antigen. When an antibody attaches to an antigen, it helps the body destroy or inactivate the antigen.
- **Antigen**—a substance or molecule that is recognized by the immune system. The molecule can be from foreign material such as bacteria or viruses.
- **Antiserum**—a serum rich in antibodies against a particular microbe.
- **Appendix**—lymphoid organ in the intestine.
- **Autoantibodies**—antibodies that react against a person's own tissue.
- **B cells**—small white blood cells crucial to the immune defenses. Also known as B lymphocytes, they come from bone marrow and develop into blood cells called plasma cells, which are the source of antibodies.
- **Basophils**—white blood cells that contribute to inflammatory reactions. Along with mast cells, basophils are responsible for the symptoms of allergy.
- **Biological response modifiers**—substances, either natural or synthesized, that boost, direct, or restore normal immune defenses. They include interferons, interleukins, thymus hormones, and monoclonal antibodies.
- **Bone marrow**—soft tissue located in the cavities of the bones. Bone marrow is the source of all blood cells.
- **Chemokines**—certain proteins that stimulate both specific and general immune cells and help coordinate immune responses and inflammation.
- **Complement**—a complex series of blood proteins whose action "complements" the work of antibodies. Complement destroys bacteria, produces inflammation, and regulates immune reactions.
- **Complement cascade**—a precise sequence of events, usually triggered by antigen-antibody complexes, in which each component of the complement system is activated in turn.
- **Cytokines**—powerful chemical substances secreted by cells that enable the body's cells to communicate with one another. Cytokines include lymphokines produced by lymphocytes and monokines produced by monocytes and macrophages.
- **Cytotoxic T lymphocytes** (CTLs)—a subset of T cells that carry the CDS marker and can destroy body cells infected by viruses or transformed by cancer.
- **DNA (deoxyribonucleic acid)**—a long molecule found in the cell nucleus; it carries the cell's genetic information.
- **Eosinophils**—white blood cells that contain granules filled with chemicals damaging to parasites, and enzymes that affect inflammatory reactions.
- **Granules**—membrane-bound organelles within cells where proteins are store before secretion.
- **Granulocytes**—phagocytic white blood cells filled with granules; neutrophils, eosinophils, basophils, and mast cells are examples of granulocytes.
- **Growth factors**—chemicals secreted by cells that stimulate proliferation of or changes in the physical properties of other cells.

- **Helper T cells**—a subset of T cells that carry the CD4 surface marker and are essential for turning on antibody production, activating cytotoxic T cells, and initiating many other immune functions.
- **Immune response**—reaction of the immune system to foreign substances.
- **Immunoglobulins**—a family of large protein molecules, also known as antibodies, produced by B cells.
- **Immunosuppressive**—capable of reducing immune response.
- **Interferons**—proteins produced by cells that stimulate antivirus immune response or alter the physical properties of immune cells.
- **Interleukins**—a major group of lymphokines and monokines.
- **Leukocytes**—all white blood cells.
- **Lymph**—a transparent, slightly yellow fluid that carries lymphocytes, bathes the body tissues, and drains into the lymphatic vessels.
- **Lymph nodes**—small bean-shaped organs of the immune system, distributed widely throughout the body and linked by lymphatic vessels. Lymph nodes are garrisons of B, T, and other immune cells.
- **Lymphatic vessels**—a body-wide network of channels, similar to the blood vessels, which transport lymph to the immune organs and into the bloodstream.
- **Lymphocytes**—small white blood cells produced in the lymphoid organs and paramount in the immune defenses; B cells and T cells are lymphocytes.
- **Lymphoid organs**—the organs of the immune system, where lymphocytes develop and congregate. They include the bone marrow, thymus, lymph nodes, spleen, and various other clusters of lymphoid tissue.
- **Lymphokines**—powerful chemical substances secreted by lymphocytes. These molecules help direct and regulate the immune response.
- **Macrophage**—a large and versatile immune cell that devours invading pathogens and other intruders. Macrophages stimulate other immune cells by presenting them with small pieces of the invaders.
- **Mast cell**—a granulocyte found in tissue. The contents of mast cells, along with those of basophils, are responsible for allergy symptoms
- **Memory cells**—a subset of T cells and B cells that have been exposed to antigens and can then respond more readily when the immune system encounters those same antigens again.
- **Monoclonal antibodies**—antibodies produced by a single cell or its identical progeny, specific for a given antigen. As tools for binding to specific protein molecules, they are invaluable in research, medicine, and industry.
- **Monocytes**—large phagocytic white blood cells which, when entering tissue, develop into macrophages.
- **Monokines**—powerful chemical substances secreted by monocytes and macrophages. These molecules help direct and regulate the immune responses.
- **Natural killer (NK) cells**—large granule-containing lymphocytes that recognize and kill cells lacking self antigens. Their target recognition molecules are different from T cells.
- **Neutrophil**—white blood cell that is an abundant and important phagocyte.

- **Passive immunity**—immunity resulting from the transfer of antibodies or antiserum produced by another individual.
- **Phagocytes**—large white blood cells that contribute to the immune defenses by ingesting microbe or other cells and foreign particles.
- **Phagocytosis**—process by which one cell engulfs another cell or large particle.
- **Plasma cells**—large antibody-producing cells that develop from B cells.
- **Serum**—the clear liquid that separates from the blood when it is allowed to clot. This fluid contains the antibodies that were present in the whole blood.
- **Spleen**—a lymphoid organ in the abdominal cavity that is an important center for immune system activities.
- **Stem cells**—immature cells from which all cells derive. The bone marrow is rich in stem cells, which become specialized blood cells.
- **T cells**—small white blood cells that recognize antigen fragments bound to cell surfaces by specialized antibody-like receptors. "T" stands for thymus, where T cells acquire their receptors.
- **Thymus**—a primary lymphoid organ, high in the chest, where T lymphocytes proliferate and mature.
- **Tolerance**—a state of immune nonresponsiveness to a particular antigen or group of antigens.
- **Tonsils and adenoids**—prominent oval masses of lymphoid tissues on either side of the throat.

✓   **Important Point**: Lymph is the fluid in the lymphatic vessels. It is picked up from the interstitial fluid and returned to the blood plasma.

INDIGENOUS AND FOREIGN CELLS

The key to a healthy immune system is its remarkable ability to distinguish between the body's indigenous and foreign cells. The body's immune defenses normally coexist peacefully with cells that carry distinctive indigenous marker molecules. But when immune defenders encounter cells or organisms carrying markers that say, "foreign," they quickly launch an attack.

Anything that can trigger this immune response is called an *antigen*. An antigen can be a microbe such as a virus, or even a part of a microbe. Tissues or cells from another person (except an identical twin) also carry non-self markers and act as antigens. This explains why tissue transplants may be rejected.

In abnormal situations, the immune system can mistake indigenous for foreign and launch an attack against the body's own cells or tissues. The result is called an *autoimmune disease*. Some forms of arthritis and diabetes are autoimmune diseases. In other cases, the immune system responds to a seemingly harmless foreign substance such as ragweed pollen. The result is an allergy, and this kind of antigen is called an *allergen*.

Structure of the Immune System

The organs of the immune system are positioned throughout the body. They are called *lymphoid organs* because they are home to *lymphocytes*, small white blood cells that are the key players in the immune system.

Bone marrow, the soft tissue in the hollow center of bones, is the ultimate source of all blood cells, including white blood cells destined to become immune cells. The *thymus* is an organ that lies behind the breastbone; lymphocytes known as *T lymphocytes*, or just *"T cells,"* mature in the thymus.

Lymphocytes can travel throughout the body using the blood vessels. The cells can also travel through a system of lymphatic vessels that closely parallels the body's veins and arteries. Cells and fluids are exchanged between blood and lymphatic vessels, enabling the lymphatic system to monitor the body for invading microbes. The lymphatic vessels carry *lymph*, a clear fluid that bathes the body's tissues.

Small, bean-shaped *lymph nodes* are laced along the lymphatic vessels, with clusters in the neck, armpits, abdomen, and groin. Each lymph node contains specialized compartments where immune cells congregate, and where they can encounter antigens.

Immune cells and foreign particles enter the lymph nodes via incoming lymphatic vessels or the lymph nodes' tiny blood vessels. All lymphocytes exit lymph nodes through outgoing lymphatic vessels. Once in the blood stream, they are transported to tissues throughout the body. They patrol everywhere for foreign antigens, then gradually drift back into the lymphatic system, to begin the cycle all over again.

The *spleen* is a flattened organ at the upper left of the abdomen. Like the lymph nodes, the spleen contains specialized compartments where immune cells gather and work, and serves as a meeting ground where immune defenses confront antigens.

Clumps of lymphoid tissue are found in many parts of the body, especially in the linings of the digestive tract and the airways and lungs—territories that serve as gateways to the body. These tissues include the *tonsils, adenoids,* and *appendix*.

✓   **Important Point**: Lymphatic vessels carry fluid away from the tissues.

Immune Cells

The immune system stockpiles a huge arsenal of cells, not only lymphocytes but also cell-devouring *phagocytes* and their relatives. Some immune cells take on all comers, while others are trained on highly specific targets. To work effectively, most immune cells need the cooperation of their comrades. Sometimes immune cells communicate by direct physical contact, sometimes by releasing chemical messengers.

The immune system stores just a few of each kind of the different cells needed to recognize millions of possible enemies. When an antigen appears, those few matching cells multiply into a full-scale army. After their job is done, they fade away, leaving sentries behind to watch for future attacks.

All immune cells begin as immature stem cells in the bone marrow. They respond to different cytokines and other signals to grow into specific immune cell types, such as T cells, B cells, or phagocytes. Because stem cells have not yet committed to a particular future, they are an interesting possibility for treating some immune system disorders. Researchers currently are investigating if a person's own stem cells can be used to regenerate damaged immune responses in autoimmune diseases and immune deficiency diseases.

✓   **Important Point**: The right lymphatic duct drains lymph from the upper right quadrant of the body and the thoracic duct drains all the rest.

B LYMPHOCYTES

B cells, one of the main types of lymphocytes, work chiefly by secreting substances called *antibodies* into the body's fluids. Antibodies ambush antigens circulating the bloodstream. They are powerless, however, to penetrate cells. The job of attacking target cells—either cells that have been infected by viruses or cells that have been destroyed by cancer—is left to T cells or other immune cells (described below).

Each B cell is programmed to make one specific antibody. For example, one B cell will make an antibody that blocks a virus that causes the common cold, while another produces an antibody that attacks a bacterium that causes pneumonia.

When a B cell encounters its triggering antigen, it gives rise to large cells known as plasma cells. Every plasma cell is essentially a factory for producing an antibody. Each of the plasma cells descended from a given B cell manufactures millions of identical antibody molecules and pours them into the bloodstream.

An antigen matches an antibody much as a key matches a lock. Some match exactly; others fit more like a skeleton key. But whenever antigen and antibody interlock, the antibody marks the antigen for destruction.

✓   **Important Point**: Pressure gradients that move fluid through the lymphatic vessels come from the skeletal muscle action, respiratory movements, and contraction of smooth muscle in vessel walls.

Antibodies belong to a family of large molecules known as *immunoglobulins*. Different types play different roles in the immune defense strategy:

- **Immunoglobulin G, or IgG**—works efficiently to coat microbes, speeding their uptake by other cells in the immune system.
- **IgM**—is very effective at killing bacteria.
- **IgA**—concentrates in body fluid including tears, saliva, the secretions of the respiratory tract, and the digestive tract, thereby guarding entrances to the body.
- **IgE**—is the villain responsible for the symptoms of allergy.
- **IgD**—remains attached to B cells and plays a key role in initiating early B-cell response.

T CELLS

Unlike B cells, T cells do not recognize free-floating antigens. Rather, their surfaces contain specialized antibody-like receptors that detect fragments of antigens on the surfaces of infected or cancerous cells. T cells contribute to immune defenses in 2 major ways: some direct and regulate immune response; others directly attack infected or cancerous cells.

Helper T cells coordinate immune responses by communicating with other cells. Some stimulate nearby B cells to produce antibodies, others call in microbe-gobbling cells called phagocytes, still others activate other T cells.

Killer T cells—also called *cytotoxic T lymphocytes* or *CTLs*—perform a different function. These cells directly attack other cells carrying certain foreign or abnormal molecules on their surfaces. CTLs are especially useful for attacking viruses because viruses often hide from other parts of the immune system while they grow inside infected cells. CTLs recognize small fragments of these viruses peeking out from the cell membrane and launch an attack to kill the cell.

In most cases, T cells only recognize an antigen if it is carried on the surface of a cell by one of the body's own major histocompatibility complex (MHC) molecules. MHC molecules are proteins recognized by T cells when distinguishing between indigenous and foreign. An indigenous MHC molecule provides a recognizable scaffolding to present a foreign antigen to the T cell.

Although MHC molecules are required for T-cell responses against foreign invaders, they also pose a difficulty during organ transplantations. Virtually every cell in the body is covered with MHC proteins, but each person has a different set of these proteins on his or her cells. If a T cell recognizes a foreign MHC molecule on another cell, it will destroy the cell. Therefore, doctors must match organ recipients with donors who have the closest MHC makeup. Otherwise the recipient's T cells will likely attack the transplanted organ, leading to *graft rejection*.

Natural killer (NK) cells are another kind of lethal white cell, or lymphocyte. Like killer T cells, NK cells are armed with *granules* filled with potent chemicals. But while killer T cells look for antigen fragments bound to self-MHC molecules, NK cells recognize cells lacking self-MHC molecules. Thus NK cells have the potential to attack many types of foreign cells.

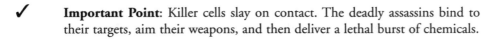 **Important Point**: Killer cells slay on contact. The deadly assassins bind to their targets, aim their weapons, and then deliver a lethal burst of chemicals.

PHAGOCYTES

Phagocytes are large white cells that can swallow and digest microbes and other foreign particles. *Monocytes* are phagocytes that circulate in the blood. When monocytes migrate into tissues, they develop into *macrophages*. Specialized types of macrophages can be found in many organs, including lungs, kidneys, brain, and liver.

Macrophages play many roles. As scavengers, they rid the body of worn-out cells and other debris. They display bits of foreign antigen in a way that draws the attention of matching lymphocytes. And they churn out an amazing variety of powerful chemical signals, known as *monokines*, which are vital to the immune responses.

Granulocytes are another kind of immune cell. They contain granules filled with potent chemicals, which allow the granulocytes to destroy microorganisms. Some of these chemicals, such as histamine, also contribute to inflammation and allergy.

One type of granulocyte, the *neutrophil*, is also a phagocyte; it uses its prepackaged chemicals to break down the microbes it ingests. *Eosinophils* and *basophils* are granulocytes that "degranulate," spraying their chemicals onto harmful cells or microbes nearby.

The *mast cell* is a twin of the basophil, except that it is not a blood cell. Rather, it is found in the lungs, skin, tongue, and linings of the nose and intestinal tract, where it is responsible for the symptoms of allergy.

A related structure, the blood *platelet*, is a cell fragment. Platelets, too, contain granules. In addition to promoting blood clotting and wound repair, platelets activate some of the immune defenses.

✓ **Important Point**: Lymph enters a lymph node through afferent vessels, filters through the sinuses, and leaves through efferent vessels.

CYTOKINES

Components of the immune system communicate with one another by exchanging chemical messengers called cytokines. These proteins are secreted by cells and act on other cells to coordinate an appropriate immune response. Cytokines include a diverse assortment of *interleukins, interferons*, and *growth factors*.

✓ **Interesting Point**: Some cytokines are chemical switches that turn certain immune cell types on and off.

One cytokine, interleukin 2 (Il-2), triggers the immune system to produce T cells. Il-2's immunity-boosting properties have traditionally made it a promising treatment for several illnesses. Clinical studies are ongoing to test its benefits in other diseases such as cancer, hepatitis C, and HIV infection and AIDS.

Other cytokines chemically attract specific cell types. These so-called *chemokines* are released by cells at a site of injury or infection and call other immune cells to the region to help repair the damage or fight off the invader. Chemokines often play a key role in inflammation and are a promising target for new drugs to help regulated immune responses.

✓ **Important Point**: Tonsils are clusters of lymphatic tissue associated with openings into the pharynx. They provide protection against pathogens that may enter through the nose and mouth.

COMPLEMENT

The *complement* system is made up of about 25 proteins that work together to "complement" the action of antibodies in destroying bacteria. Complement also helps to rid the body of antibody-coated antigens. Complement proteins, which cause blood vessels to become dilated and then leaky, contribute to the redness, warmth, swelling, pain, and loss of function that characterize an *inflammatory response*.

Complement proteins circulate in the blood in an inactive form. When the first protein in the complement series is activated—typically by antibody that has locked onto an antigen—it sets in motion a domino effect. Each component takes its turn in a precise chain of steps known as the complement cascade. The end product is a cylinder inserted into—and puncturing a hole in—the cell's wall. With fluids and molecules flowing in and out, the cell swells and bursts. Other components of the complement system make bacteria more susceptible to phagocytosis or beckon other cells to the area.

✓   **Important Point**: The spleen is a lymph organ that filters blood and also acts as a reservoir for blood.

Immune Response

Infections are the most common cause of human disease. They range from the common cold to debilitating conditions like chronic hepatitis to life-threatening diseases such as AIDS. Disease-causing microbes (*pathogens*) attempting to get into the body must first move past the body's external armor, usually the skin or cells lining the body's internal passageways.

The skin provides an imposing barrier to invading microbes. It is generally penetrable only through cuts or tiny abrasions. The digestive and respiratory tracts—both portals of entry for a number of microbes—also have their own levels of protection. Microbes entering the nose often cause the nasal surfaces to secrete more protective mucus, and attempts to enter the nose or lungs can trigger a sneeze or cough reflex to force microbial invaders out of the respiratory passageways. The stomach contains a strong acid that destroys many pathogens that are swallowed with food.

If microbes survive the body's front-line defenses, they still have to find a way through the walls of the digestive, respiratory, or urogenital passageways to the underlying cells. These passageways are lined with tightly packed epithelial cells covered in a layer of mucus, effectively blocking the transport of many organisms. Mucosal surfaces also secrete a special class of antibody called IgA, which in many cases is the first type of antibody to encounter an invading microbe. Underneath the epithelial layer a number of cells, including macrophages, B cells, and T cells, lie in wait for any germ that might bypass the barriers at the surface.

Next, invaders must escape a series of general defenses, which are ready to attack, without regard for specific antigen markers. These include patrolling phagocytes, NK cells, and complement. Microbes that cross the general barriers then confront specific weapons tailored just for them. Specific weapons, which include both antibodies and T cells, are equipped with singular receptor structures that allow them to recognize and interact with their designated targets.

✓   **Important Point**: The thymus is large in the infant and atrophies after puberty.

BACTERIA, VIRUSES, AND PARASITES

The most common disease-causing microbes are bacteria, viruses, and parasites. Each uses a different tactic to infect a person, and, therefore, each is thwarted by a different part of the immune system.

Most bacteria live in the spaces between cells and are readily attacked by antibodies. When antibodies attach to a bacterium, they send signals to complement proteins and phagocytic cells to destroy the bound microbes. Some bacteria are eaten directly by phagocytes, which signal to certain T cells to join the attack.

All viruses, plus a few types of bacteria and parasites, must enter cells to survive, requiring a different approach. Infected cells use their MHC molecules to put pieces of the invading microbes on the cell's surface, flagging down cytotoxic T lymphocytes to destroy the infected cell. Antibodies also can assist in the immune response, attaching to and clearing viruses before they have a chance to enter the cell.

Parasites live either inside or outside cells. Intracellular parasites such as the organism that causes malaria can trigger T-cell response. Extracellular parasites are often much larger than bacteria or viruses and require a much broader immune attack. Parasitic infections often trigger an inflammatory response when eosinophils, basophils, and other specialized granular cells rush to the scene and release their stores of toxic chemicals in an attempt to destroy the invader. Antibodies also play a role in this attack, attracting the granular cells to the site of infection.

NATURAL AND ACQUIRED IMMUNITY

Long ago, physicians realized that people who had recovered from the plague would never get it again—they had acquired immunity. This is because some of the activated T and B cells become *memory cells*. The next time an individual meets up with the same antigen, the immune system is set to demolish it.

✓   **Important Point**: Immunity can be strong or weak, short-lived or long-lasting, depending on the type of antigen, the amount of antigen, and the route by which it enters the body.

Immunity can also be influenced by inherited genes. When faced with the same antigen, some individuals will respond forcefully, others feebly, and some not at all.

An immune response can be sparked not only by infection but also by immunization with *vaccines*. Vaccines contain microorganisms—or parts of microorganisms—that have been treated so they can provoke an immune response but not full-blown disease.

✓   **Important Point**: Immunity can also be transferred from one individual to another by injections of *serum* rich in antibodies against a particular microbe (*antiserum*). For example, immune serum is sometimes given to protect travelers to countries where hepatitis A is widespread. Such passive immunity typically lasts only a few weeks or months.

Infants are born with weak immune responses but are protected for the first few months of life by antibodies received from their mothers before birth. Babies who are nursed can also receive some antibodies from breast milk that help to protect their digestive tracts.

IMMUNE TOLERANCE

Immune tolerance is the tendency of T or B lymphocytes to ignore the body's own tissues. Maintaining tolerance is important because it prevents the immune system from attacking its fellow cells. Scientists are hard at work trying to understand how the immune system knows when to respond and when to ignore.

Tolerance occurs in at least 2 ways. Central tolerance occurs during lymphocyte development. Very early in each immune cell's life, it is exposed to many of the self molecules in the body. If it encounters these molecules before it has fully matured, the encounter activates an internal self-destruct pathway and the immune cell dies. This process, called clonal deletion, helps ensure that self-reactive T cells and B cells do not mature and attack healthy tissues.

Because maturing lymphocytes do not encounter every molecule in the body, they must also learn to ignore mature cells and tissues. In peripheral tolerance, circulating lymphocytes might recognize an indigenous molecule but cannot respond because some of the chemical signals required to activate the T or B cell are absent. So-called *clonal anergy*, therefore, keeps potentially harmful lymphocytes switched off. Peripheral tolerance may also be imposed by a special class of regulatory T cells that inhibits helper or cytotoxic T-cell activation by indigenous antigens.

VACCINES

Medical workers have long helped the body's immune system prepare for future attacks through vaccination. Vaccines consist of killed or modified microbes, components of microbes, or microbial DNA that trick the body into thinking an infection has occurred. An immunized person's immune system attacks the harmless vaccine and prepares for subsequent invasions. Vaccines remain one of the best ways to prevent infectious diseases and have an excellent safety record. Previously devastating diseases such as smallpox, polio, and whooping cough have been greatly controlled or eliminated through worldwide vaccination programs.

Disorders of the Immune System

ALLERGIC DISEASES

The most common types of allergic diseases occur when the immune system responds to a false alarm. In an allergic person, a normally harmless material such as grass pollen or house dust is mistaken for a threat and attacked.

✓ **Important Point**: Allergies such as pollen allergy are related to the antibody known as IgE. Like other antibodies, each IgE antibody is specific; one acts against oak pollen, another against ragweed.

AUTOIMMUNE DISEASES

Sometimes the immune system's recognition apparatus breaks down, and the body begins to manufacture T cells and antibodies directed against its own cells and organs. Misguided T cells and autoantibodies, as they are known, contribute to many diseases. For instance, T cells that attack pancreas cells contribute to diabetes, while an autantibody known as rheumatoid factor is common in people with rheumatoid arthritis. People with system lupus erythematosus (SLE) have antibodies to many types of their own cells and cell components.

No one knows exactly what causes an autoimmune disease, but multiple factors are likely to be involved. These include elements in the environment, such as viruses, certain drugs, and sunlight, all of which may damage or alter normal body cells. Hormones are suspected of playing a role, since most autoimmune diseases are far more common in women than in men. Heredity, too, seems to be important. Many people with autoimmune disease have characteristic types of self marker molecules.

IMMUNE COMPLEX DISEASES

Immune complexes are clusters of interlocking antigens and antibodies. Normally, immune complexes are rapidly removed from the bloodstream. Sometimes, however, they continue to circulate, and eventually become trapped in the tissues of the kidneys, the lungs, skin, joints, or blood vessels. There they set off reactions with complement that lead to inflammation and tissue damage.

✓ **Important Point**: Immune complexes work their mischief in many diseases. These include malaria and viral hepatitis, as well as many autoimmune diseases.

IMMUNODEFICIENCY DISORDERS

When the immune system is missing one or more of its components, the result is an immunodeficiency disorder. Immunodeficiency disorders can be inherited, acquired through infection, or produced unintentionally by drugs such as those used to treat people with cancer or those who have received transplants.

Temporary immune deficiencies can develop in the wake of common virus infections, including influenza, infectious mononucleosis, and measles. Immune responses can also be depressed by blood transfusions, surgery, malnutrition, smoking, and stress.

Some children are born with poorly functioning immune systems. Some have flaws in the B cell system and cannot produce antibodies. Others, whose thymus is either missing or small and abnormal, lack T cells. Very rarely, infants are born lacking all of the major immune defenses. This condition is known as severe combined immunodeficiency disease (SCID).

AIDS is an immunodeficiency disorder caused by a virus (HIV) that infects immune cells. HIV can destroy or disable vital T cells, paving the way for a variety of immunologic shortcomings. HIV also can hide out for long periods in immune cells. As the immune defenses falter, a person with AIDS falls prey to unusual, often life-threatening infections and rare cancers.

A contagious disease, AIDS is spread by intimate sexual contact, transfer of the virus from mother to infant during pregnancy, or direct blood contamination. There is no cure for AIDS, but newly developed antiviral drugs can slow the advance of the disease, at least for a time.

✓ **Important Point**: Edema is the swelling that results from inadequate drainage of lymph from the body tissue spaces. Edema is brought about by heart and kidney disorders, malnutrition, injury, or other causes.

Chapter Review Questions

1. The _____ are glands that filter bacteria from the tissue fluids.
2. _____ are white blood cells that participate in immune responses.
3. The _____ is our primary defense mechanism against pathogens.
4. _____ are a subset of T cells that carry the CD4 surface marker.
5. They become macrophages upon entering tissue: _____.
6. _____ is the soft tissue in the hollow center of bones.
7. All immune cells begin as _____ in the bone marrow.
8. _____ ambush antigens circulating the bloodstream.
9. _____ can swallow and digest microbes.
10. Twin of the basophil: _____.
11. Made up of 25 proteins: _____.
12. Keeps potentially harmful lymphocytes switched off: _____.

References and Additional Reading

Braem, T. 1994. *The Lymphatic System*. Bryan Edwards Publishing.

Janeway, C. 2004. *Immunobiology*, 6th ed. New York: Garland Science.

NIH. 2003. *Understanding the Immune System: How It Works*. Washington, DC: National Institutes of Health.

Parham, P. 2004. *The Immune System*, 2nd ed. New York: Garland Science.

SEER. 2000. *Anatomy and Physiology: Functions of the Lymphatic System*. Accessed August 19, 2006, at /www.training.seer.concer.gov/.

Sompayrac, L.M. 2002. *How the Immune System Works,* 2nd ed. Oxford: Blackwell Science, Inc.

CHAPTER 11

Digestive System

In a human's lifetime, the digestive system may handle about 50 tons of food!

Topics in This Chapter

- Digestion Process
- Control of the Digestive Process

When we ingest food and drink, unless it is quite tasty, filling and/or refreshing, we give the process little thought—or none at all. Even when it comes time to discharge those items we have ingested or drunk into the toilet or urinal, little thought beyond that natural urge crosses our minds.

Some folks live to eat and others eat to live. No matter the circumstance or your point of view on eating and drinking, when we ingest or drink, the life- and health-sustaining process that we hardly ever think about is just beginning. When we eat such things as meat, bread, crackers, licorice, donuts, honeybuns, ice cream, and vegetables, they are not in a form that the body can use as nourishment. Our food and drink must be changed into smaller molecules of nutrients before they can be absorbed into the blood and carried to cells throughout the body. **Digestion** is the process by which food and drink are broken down into their smallest parts so that the body can use them to build and nourish cells and to provide energy.

The human digestive system, commonly called the *gut, gastrointestinal tract*, or *alimentary canal*, is somewhat similar to the earthworm's in basic design, though it is more complicated and efficient. The stomach is a series of hollow organs joined in a long, twisting tube from the mouth to the anus. Inside the tube is a lining called the mucosa. In the mouth, stomach, and small intestine, the mucosa contains tiny

---

Note: Much of the material in this chapter is from NIH's 2004 *Your Digestive System and How It Works*, accessible at http://digestive.niddk.nih.gov.

glands that produce juices to help digest food. Two solid organs, the liver and the pancreas, produce digestive juices that reach the intestine through small tubes. In addition, parts of other organ systems (e.g., nerves and blood) play a major role in the digestive system.

Digestion Process

Before discussing the digestion process, it is important to be familiar with the key terms and definitions involved.

DEFINITION OF KEY TERMS (ENCHANTED LEARNING 2001)

- **Anus**—the opening at the end of the digestive system from which feces (waste) exits the body.
- **Appendix**—a small sac located on the cecum.
- **Ascending colon**—the part of the large intestine that run upward; it is located after the cecum.
- **Bile**—a digestive chemical that is produced in the liver, stored in the gall bladder, and secreted into the small intestine.
- **Cecum**—the first part of the large intestine; the appendix is connected to the cecum.
- **Chyme**—food in the stomach that is partly digested and mixed with stomach acids. Chyme goes on to small intestine for further digestion.
- **Descending colon**—the part of the large intestine that run downward after the transverse colon and before the sigmoid colon.
- **Duodenum**—the first part of the small intestine; it is C-shaped and runs from the stomach to the jejunum.
- **Epiglottis**—the flap at the back of the tongue that keeps chewed food from going down the windpipe to the lungs. When you swallow, the epiglottis automatically closes. When you breathe, the epiglottis opens so that air can go in and out of the windpipe.
- **Esophagus**—the long tube between the mouth and the stomach. It uses rhythmic muscle movements (called peristalsis) to force food from the throat into the stomach.
- **Gall bladder**—a small, sac-like organ located by the duodenum. It stores and releases bile (a digestive chemical which is produced in the liver) into the small intestine.
- **Ileum**—the last pat of the small intestine before the large intestine begins.
- **Jejunum**—the long, coiled mid-section of the small intestine; it is between the duodenum and the ileum.
- **Liver**—a large organ located above and in front of the stomach. It filters toxins from the blood, and makes bile (which breaks down fats) and some blood proteins.
- **Mouth**—the first part of the digestive system, where food enters the body. Chewing and salivary enzymes in the mouth are the beginning of the digestive process (breaking down the food).

- **Pancreas**—an enzyme-producing gland located below the stomach and above the intestines. Enzymes from the pancreas help in the digestion of carbohydrates, fats and proteins in the small intestine.
- **Peristalsis**—rhythmic muscle movements that force food in the esophagus from the throat into the stomach. Peristalsis is involuntary—you cannot control it. It is also what allows you to eat and drink when upside-down.
- **Rectum**—the lower part of the large intestine, where feces are stored before they are excreted.
- **Salivary glands**—glands located in the mouth that produce saliva. Saliva contains enzymes that break down carbohydrates (starch) into smaller molecules.
- **Sigmoid colon**—the part of the large intestine between the descending colon and the rectum.
- **Stomach**—a sac-like, muscular organ that is attached to the esophagus. Both chemical and mechanical digestion take place in the stomach. When food enters the stomach, it is churned in a bath of acids and enzymes.
- **Transverse colon**—the part of the large intestine that runs horizontally across the abdomen.

Simply, digestion involves the mixing of food, its movement through the digestive tract, and the chemical breakdown of the large molecules of food into smaller molecules. Digestion begins in the mouth when we chew and swallow, and is completed in the small intestine. The chemical process varies somewhat for different kinds of food.

While in the mouth and during chewing, food is broken down by the chemical action of salivary enzymes (these enzymes are produced by the salivary glands and break down starches into smaller molecules).

The large, hollow organs of the digestive system contain muscle that enables their walls to move. The movement of organ walls can propel food and liquid and also can mix the contents with each organ. Typical movement of the esophagus, stomach, and intestine is called *peristalsis*. The action of peristalsis looks like an ocean wave moving through the muscle. The muscle of the organ produces a narrowing and then propels the narrowed portion slowly down the length of the organ. These waves of narrowing push the food and fluid in front of them through each hollow organ.

The first major muscle movement occurs when food or liquid is swallowed. Although we are able to start swallowing by choice, once the swallow begins, it becomes involuntary and proceeds under the control of the nerves.

✓   **Fact or Fallacy**: Spicy food and stress cause stomach ulcers.

False
The truth is, almost all stomach ulcers are caused either by infection with a bacterium called *Helicobacter pylori (H. pylori)* or by use of pain medications such as aspirin, ibuprofen, or naproxen, the so-called nonsteroidal anti-inflammatory drugs (NSAIDs). Most *H. pylori*-related ulcers can be cured with antibiotics. NSAID-induced ulcers can be cured with time, stomach-protective medications, antacids, and avoidance of NSAIDs. Spicy food and

stress may aggravate ulcer symptoms in some people, but they do not cause ulcers. Ulcers can also be caused by cancer (NIH 2003).

The *esophagus* is the organ into which the swallowed food is pushed. It connects the throat above with the stomach below. At the junction of the esophagus and stomach, there is a ringlike valve (sphincter) closing the passage between the 2 organs. However, as the food approaches the closed ring, the surrounding muscles relax and allow the food to pass.

The food then enters the stomach, which has 3 mechanical tasks to do. First, the stomach must store the swallowed food and liquid. This requires the muscle of the upper part of the stomach to relax and accept large volumes of swallowed material. The second job is to mix up the food, liquid, and digestive juice produced by the stomach. The lower part of the stomach mixes these materials by its muscle action. The third task of the stomach is to empty it contents slowly into the small intestine.

✓   **Interesting Point**: An expanded stomach can hold approximately 2 liters (approx. 0.528 gals) of food.

Several factors affect emptying of the stomach, including the nature of the food (mainly its fat and protein content) and the degree of muscle action of the emptying stomach and the next organ to receive the contents (the small intestine). As the food is digested in the small intestine and dissolved into the juices from the pancreas and liver, the contents of the intestine are mixed and pushed forward to allow further digestion.

Finally, all of the digested nutrients are absorbed through the intestinal walls. The waste products of this process include undigested parts of the food, known as fiber, and older cells that have been shed from the mucosa. These materials are propelled into the colon, where they remain, usually for a day or two, until the feces are expelled by a bowel movement.

✓   **Fact or Fallacy**: Bowel regularity means a bowel movement every day.

False
The frequency of bowel movements among normal, healthy people varies from 3 a day to 3 a week and some perfectly healthy people fall outside both ends of this range (NIH 2003).

As mentioned, the glands that act first are in the mouth—the *salivary glands*. Saliva produced by these glands contains an enzyme that begins to digest the starch from food into smaller molecules. The next set of digestive glands is in the stomach lining. They produce stomach acid and an enzyme that digests protein. One of the unsolved puzzles of the digestive system is why the acid juice of the stomach does not dissolve the tissue of the stomach itself. In most people, the stomach mucosa is able to resist this acidic juice, although food and other tissues of the body cannot.

After the stomach empties the food and juice mixture into the small intestine, the juices of 2 other digestive organs mix with the food to continue the process of digestion. One of these organs is the pancreas. It produces a juice that contains a wide array of enzymes to break down the carbohydrate, fat, and protein in food. Other enzymes that are active in the process come from glands in the wall of the intestine or even a part of that wall.

The liver produces yet another digestive juice—*bile*. The bile is stored between meals in the gallbladder. At mealtime, it is squeezed out of the gall bladder into the bile ducts to reach the intestine and mix with the fat in our food. The bile acids dissolve the fat into the watery contents of the intestine, much like detergents that dissolve grease from a frying pan. After the fat is dissolved, it is digested by enzymes from the pancreas and the lining of the intestine.

✓   **Fact or Fallacy**: Habitual use of enemas to treat constipation is harmless.

False
Habitual use of enemas is not harmless. Over time, enemas can impair the natural muscle action of the intestine, leaving them unable to function normally. An ongoing need for enemas is not normal; you should see a doctor if you find yourself relying on them or any other medication to have a bowel movement (NIH 2003).

Digested molecules of food, as well as water and minerals from the diet, are absorbed from the cavity of the upper small intestine. Most absorbed materials cross the mucosa into the blood and are carried off in the bloodstream to other parts of the body for storage or further chemical change. As already noted, this part of the process varies with different types of nutrients:

• **Carbohydrates**—It is recommended that about 55%–60% of total daily calories be from carbohydrates. Some of our most common foods contain mostly carbohydrates. Examples are bread, potatoes, legumes, rice, spaghetti, fruits, and vegetables. Many of these foods contain both starch and fiber.

The digestive carbohydrates are broken into simpler molecules by enzymes in the saliva, in juice produced by the pancreas, and in the lining of the small intestine. Starch is digested in 2 steps: First, enzymes in the saliva and pancreatic juice break the starch into molecules called maltose; then an enzyme in the lining of the small intestine (maltase) splits the maltose into glucose molecules that can be absorbed into the blood. Glucose is carried through the bloodstream to the liver, where it is stored or used to provide energy for the work of the body.

Table sugar is another carbohydrate that must be digested to be useful. An enzyme in the lining of the small intestine digests table sugar into glucose and fructose, each of which can be absorbed from the intestinal cavity into the blood. Milk contains yet another type of sugar, lactose, which is changed into absorbable molecules by an enzyme called lactase, also found in the intestinal lining.

- **Protein**—Foods such as meat, eggs, and beans consist of giant molecules of protein that must be digested by enzymes before they can be used to build and repair body tissues. An enzyme in the juice of the stomach starts the digestion of swallowed protein. Further digestion of the protein is completed in the small intestine. Here, several enzymes from the pancreatic juice and the lining of the intestine carry out the breakdown of huge protein molecules into small molecules called amino acids. These small molecules can be absorbed from the hollow of the small intestine into the blood and then be carried to all parts of the body to build the walls and other parts of cells.
- **Fats**—Fat molecules are a rich source of energy for the body. The first step in digestion of a fat such as butter is to dissolve it into the watery content of the intestinal cavity. The bile acids produced by the liver act as natural detergents to dissolve fat in water and allow the enzymes to break the large fat molecules into smaller molecules, some of which are fatty acids and cholesterol. The bile acids combine with the fatty acids and cholesterol and help these molecules to move into the cells of the mucosa. In these cells the small molecules are formed back into large molecules, most of which pass into vessels (called lymphatics) near the intestine. These small vessels carry the reformed fat to the veins of the chest, and the blood carries the fat to storage depots in different parts of the body.
- **Vitamins**—Another vital part of our food that is absorbed from the small intestine is the class of chemicals we call vitamins. The 2 different types of vitamins are classified by the fluid in which they can be dissolved: water-soluble vitamins (all the B vitamins and vitamin C) and fat-soluble vitamins (vitamins A, D, and K).
- **Water and salt**—Most of the material absorbed from the cavity of the small intestine is water in which salt is dissolved. The salt and water come from the food and liquid we swallow and the juices secreted by the many digestive glands.

✓   **Fact or Fallacy**: Irritable bowel syndrome is a disease.

True
Irritable bowel syndrome is a disease, although it is also called a "functional disorder." Irritable bowel syndrome involves a problem in how the muscles in the intestines work, and pain perception on the bowel. It is characterized by gas, abdominal pain, and diarrhea or constipation, or both. Although the syndrome can cause considerable pain and discomfort, it does not damage the digestive tract as organic diseases do. Also, irritable bowel syndrome does not lead to more serious digestive diseases later, such as cancer (NIH 2003).

Control of the Digestive Process

A fascinating feature of the digestive system is that it contains its own regulators. The major hormones that control the functions of the digestive system are produced and released by cells in the mucosa of the stomach and small intestine. These hormones are released into the blood of the digestive tract, travel back to the heart and through the arteries, and return to the digestive system, where they stimulate digestive juices and cause organ movement.

The hormones that control digestion are gastrin, secretin, and cholecystokinin (CCK):

- **Gastrin** causes the stomach to produce an acid for dissolving and digesting some foods. It is also necessary for the normal growth of the lining of the stomach, small intestine, and colon.
- **Secretin** causes the pancreas to send out a digestive juice that is rich in bicarbonate. It stimulates the stomach to produce pepsin, an enzyme that digests protein, and it also stimulates the liver to produce bile.
- **CCK** causes the pancreas to grow and to produce the enzymes of pancreatic juice, and it causes the gallbladder to empty.

Additional hormones in the digestive system regulate appetite:

- **Ghrelin** is produced in the stomach and upper intestine in the absence of food in the digestive system and stimulates appetite.
- **Peptide YY** is produced in the GI tract in response to a meal in the system and inhibits appetite.

Both of these hormones work on the brain to help regulate the intake of food for energy.

✓   **Fact or Fallacy**: Diverticulosis is a serious but uncommon problem.

False
Actually, the majority of Americans over age 60 have diverticulosis, but only a small percentage has symptoms or complications. Diverticulosis is a condition in which little sacs or out-pouchings called diverticula develop in the wall of the colon. These sacs tend to appear and increase in number with age. Most people have no symptoms and learn that they have diverticula after an x-ray or intestinal examination. Less than 10% of people with diverticulosis ever develop complications such as infection (diverticulitis), bleeding, or perforation of the colon (NIH 2003).

Two types of nerves help to control the action of the digestive system. Extrinsic (outside) nerves come to the digestive organs from the unconscious part of the brain or from the spinal cord. They release a chemical called acetylcholine and another called adrenaline. Acetylcholine causes the muscles of the digestive organs to squeeze with more force and increase the "push" of food and juice through the digestive tract. Acetylcholine also causes the stomach and pancreas to produce more digestive juice. Adrenaline relaxes the muscle of the stomach and intestine and decreases the flow of blood to theses organs.

Even more important, though, are the intrinsic (inside) nerves, which make up a very dense network embedded in the walls of the esophagus, stomach, small intestine, and colon. The intrinsic nerves are triggered to act when the walls of the hollow organs are stretched by food. They release many different substances that speed up or delay the movement of food and the production of juices by the digestive organs.

Chapter Review Questions

1. Digestion begins in the _____.
2. The major work of digestion occurs in the _____.
3. The _____ is the widest organ in the alimentary canal.
4. Two glands lying outside the alimentary canal but important to digestion are the _____ and _____.
5. The _____ synthesizes bile.
6. _____ molecules are absorbed into villi and then enter the bloodstream through the thoracic duct.
7. The _____ is the first part of the large intestine.
8. Part of large intestine between the descending colon and the rectum: _____.
9. Digests starch from food in the mouth: _____.
10. Milk sugar: _____.
11. _____ is necessary for the normal growth of the lining of the stomach.
12. _____ inhibits appetite.

References and Additional Reading

Enchanted Learning. 2001. *Human Digestive System.* Accessible at www.Enchantedlearning.com.

Lipski, E. 1999. *Digestive Wellness.* New York: McGraw-Hill.

King, J.E. 2004. *Mayo Clinic on Digestive Health*, 2nd ed. Rochester, MN: Mayo Clinic.

Nichols, T.W. 2004. *Optimal Digestive Health: A Complete Guide.* New Page Guide.

NIH. 2003. *Facts and Fallacies about Digestive Diseases.* Washington, DC: National Digestive Diseases Information Clearinghouse. NIH Publication No. 04-2673.

NIH. 2004. *Your Digestive System and How It Works.* Washington, DC: National Digestive Diseases Information Clearinghouse. NIH Publication No. 04–2681.

Sovokie, A.M. 2004. *Gut Wisdom: Understanding and Improving Your Digestive Health.* New Page Books.

CHAPTER 12

Nervous System

The development of the brain and the nervous system is what makes humans significantly different from other animals.

Topics in This Chapter

- Organization of Nervous System
- Cells of Nervous System
- Disorders of the Nervous System

The human nervous system is the complex control center for the human body. The nervous system works rapidly by transmitting electrochemical impulses. It orchestrates body functions to maintain homoeostasis—all sensory and motor activities are integrated to achieve the balance of homeostasis. It detects change in and around the body. It processes the incoming sensory information and generates an appropriate motor response to adjust activity of muscles and glands.

The nervous system is made up of 2 major parts: the **central nervous system (CNS)** and the **peripheral nervous system (PNS)**. Nervous system tissue is specialized to receive stimuli from the outside environment and to conduct impulses to other body tissues.

The basic unit of function of the nervous system is the neuron. **Neurons** are highly specialized cells responsible for the transmission of information. They transmit their messages through a series of reversal and restoration of electrical charges across the cell membranes.

DEFINITION OF KEY TERMS

Before beginning a brief description of the human nervous system, it is important to define pertinent terms:

- **Acetylcholine**—a chemical in the brain that acts as a neurotransmitter.
- **Alzheimer's disease**—a progressive, degenerative disease that occurs in the brain and results in impaired memory, thinking, and behavior.
- **Axon**—the long, hairlike extension of a nerve cell that carries a message to the next nerve cell.
- **Basal ganglia**—several large clusters of nerve cells, including the striatum and the substantia nigra, deep in the brain below the cerebral hemispheres.
- **Bradykinesia**—slowness of movement.
- **Central nervous system** (CNS)—the brain and the spinal cord.
- **Cerebellum**—a large structure consisting of 2 halves (hemispheres) located in the lower part of the brain; responsible for the coordination of movement and balance.
- **Cerebrum**—consists of 2 parts (lobes), left and right, which form the largest and most developed part of the brain; initiation and coordination of all voluntary movement take place within the cerebrum. The basal ganglia are located immediately below the cerebrum.
- **Dementia**—not a disease itself, but a group of symptoms that characterize diseases and conditions, it is commonly defined as a decline in intellectual function that is severe enough to interfere with the ability to perform routine activities.
- **Dendrite**—a threadlike extension from a nerve cell that serves as an antenna to receive messages from the axons of other nerve cells.
- **Encephalitis**—a viral infection of the brain.
- **Ependymal cells**—are specialized epithelial cells in the CNS that produces cerebrospinal fluid.
- **Ganglion**—a cluster of nerve cells.
- **Meningitis**—an inflammation of the meninges, the membranes that cover the brain.
- **Multiple sclerosis**—is a CNS disease where the myelin sheath of the motor neurons is degenerating or being destroyed, which interferes with neuronal impulses.
- **Neuron**—a cell specialized to conduct and generate electrical impulses and to carry information from one part of the brain to another.
- **Neurotransmitters**—chemical substances that carry impulses from one nerve cell to another; found in the space (synapse) that separates the transmitting neuron's terminal (axon) from the receiving neuron's terminal (dendrite).
- **Nissl substance**—the dark granular substance inside neuronal cell bodies.
- **Oligodendrocytes**—are cells that electrically insulate neuronal axons in the central nervous system.
- **Parkinson's disease (PD)**—the most common form of Parkinson's is a slowly progressing, degenerative disease that is usually associated with the following symptoms, all of which result from the loss of dopamine-producing brain cells: tremor or trembling of the arms, jaw, legs, and face; stiffness or rigidity of the limbs and trunk;

bradykinesia—slowness of movement, postural instability, or impaired balance and coordination.

- **Synapse**—a tiny gap between the ends of nerve fibers across which nerve impulses pass from one neuron to another; at the synapse, an impulse causes the release of a neurotransmitter, which diffuses across the gap and triggers an electrical impulse in the next neuron.

Organization of the Nervous System

As mentioned, the nervous system is made up of 2 major parts: the central nervous system and the peripheral nervous system.

CENTRAL NERVOUS SYSTEM (CNS)

The brain and spinal cord are the organs of the central nervous system. Because they are so vitally important—they make up the control center that interprets, integrates, and issues commands (both voluntary and reflex actions) to the other branches of the nervous system—the brain and spinal cord, located in the dorsal body cavity, are encased in bone for protection. The brain is the cranial vault, and the spinal cord is the vertebral canal of the vertebral column. Fluid and tissue also insulate the brain and spinal cord area of the brain. Although considered to be 2 separate organs, the brain and spinal cord are continuous at the foramen magnum.

The **brain** is a highly complex organ made up almost entirely of interneurons, which includes the following structures:

- **Cerebrum**—the largest portion of the brain. Located in the upper skull, it is the seat of consciousness. The cerebrum controls all voluntary movement, sensory perception, speech, memory, and creative thought.
- **Cerebellum**—located at the rear of, and beneath, the cerebrum. It does not initiate voluntary movement, but it helps fine-tune it. The cerebellum makes sure that movements are coordinated and balanced.
- **Brainstem**—connects the brain with the spinal cord. Specifically, a portion of it known as the *medulla oblongata* is responsible for the control of involuntary functions such as breathing, cardiovascular regulation, and swallowing. The medulla oblongata is absolutely essential for life and processes a great deal of information. The medulla also helps maintain alertness.
- **Hypothalamus**—a structure in the midbrain responsible for the maintenance of homeostasis. It regulates temperature, controls hunger and thirst, and manages water balance. It also helps generate emotion.

The **spinal cord** consists of bundles of nerves within a tube-like organ. The cord comes down through the vertebrae, where connecting nerves branch off. It, too, is surrounded by cerebrospinal fluid circulating with *meningal membranes*. The spinal cord has 2 primary functions: (1) it relays messages from the brain to the motor nerves in the rest of the body; and (2) it returns impressions from the sensory nerves to the brain.

The spinal cord consists of the 3 types of neurons (i.e., sensory, motor, and interneurons). Axons of motor neurons extend from the spinal column into the peripheral nervous system, while the fibers of sensory neurons merge into the column from the PNS. Interneurons link the motor and sensory neurons, and they make up the majority of the neurons in the spinal column. In addition to the neurons, cells called glial cells (nonneuronal cells that provide support and nutrition) are present to provide physical and metabolic support for neurons. The spinal cord serves as a link between the body and the brain, and it can also regulate simple reflexes.

✓   **Important Point**: The brain and spinal cord are bathed in a fluid called the cerebrospinal fluid, which helps to cushion these delicate organs against damage. The cerebrospinal fluid is maintained by the glial cells.

PERIPHERAL NERVOUS SYSTEM (PNS)

The organs of the peripheral nervous system are the nerves and ganglia. Nerves are bundles of nerve fibers, much like muscles are bundles of muscle fibers. Cranial nerves and spinal nerves extend from the CNS to peripheral organs such as muscles and glands. Ganglia are collections, or small knots, of nerve cell bodies outside the CNS. The PNS consists of a **sensory system** (afferent) that carries information from the senses into the central nervous system from the body, and a **motor system** (efferent) that branches out from the CNS to targeted organs or muscles. The motor divisions can be divided into the **somatic system** and the **autonomic system**:

- **Somatic nervous system**—consists of nerves (neurons) that carry information to skeletal muscles which control *voluntary* (or conscious) movement. All of the neurons in the somatic system release acetylcholine, an excitatory neuron transmitter that causes skeletal muscles to contract.
- **Autonomic system**—consists of nerves that carry information to heart, visceral organs and glands. We do not have voluntary control over these actions. Autonomic neurons can either excite or inhibit their target muscles or organs. The autonomic nervous system can itself be subdivided into the sympathetic division and parasympathetic division. These 2 systems act antagonistically and often have opposite effects:
  1. **Sympathetic division**—prepares the body for emergency situations. That is, it helps us to get away from potentially dangerous situations (fight or flight response).
  2. **Parasympathetic division**—has the opposite effect of the sympathetic nervous system (i.e., rest, relax, and digest). It slows the heart rate, increases digestion, and slows breathing.

Cells of the Nervous System

The basic unit of function of the nervous system is the **neuron**, or nerve cell. Neurons are highly specialized cells that function to provide almost instantaneous transmission of electrochemical signals. Neurons are functionally classified as *afferent* (carry sensory impulses towards CNS), *efferent* (carry motor impulses way from CNS) and *association* (carry impulses from neuron to neuron).

Structurally, the neuron is an elongated cell that usually consists of 3 main parts: the **dendrites**, the **cell body** (soma), and the **axon**. The typical neuron contains many dendrites, which have the appearance of thin branches (to increase surface area) extending from the cell body. The cell body of the neuron contains the nucleus and organelles of the cell. The axon, which can sometimes be thousands of times longer than the rest of the neuron, is a single, long projection (microscopic in diameter) extending from the cell body. The axon usually ends in several small branches known as the axon terminals (axon terminals are separated from the dendrites by a small gap known as the **synapse**). Neurons are often connected in chains and networks, yet they never actually come in contact with one another (Sparknotes 2006).

The neuron has an electrical charge across the cell membrane. The actual electrical impulse moving through a neuron begins in the dendrites. From there, it passes through the cell body and then travels along the axon. The impulse always follows the same path from dendrite to cell body to axon. When the electrical impulse reaches the synapse at the end of the axon, it causes the release of specialized chemicals known as neurotransmitters. These neurotransmitters carry the signal across the synapse to the dendrites of the next neuron, starting the process again in the next cell. This electrical event allows a nerve to pass on signals to other nerves and is called an **action potential**. The action potential begins at one spot on the membrane, but spreads to adjacent areas of the membrane, propagating the message along the length of the cell membrane. After passage of the action potential, there is a brief period, the resting period, during which the membrane cannot be stimulated. This prevents the message from being transmitted backward along the membrane.

✓   **Important Point**: Neurons cannot directly pass an action potential from one to the next because of the synapse between them. Instead, neurons communicated across the synaptic clefts by the means of chemical signals know as neurotransmitters.

Nervous System Disorders

Under normal conditions, our nervous system functions billions of times at lightning speed. This all happens without us thinking about it, like the blinking of our eyelids or breathing. Only when there is some type of disorder or disease to our nervous system do we become aware of it. There are several different types of nervous system disorder. In this section we discuss a few of those that have the most significant impact on us.

MENINGITIS AND ENCEPHALITIS

According to the National Institute of Neurological Disorders and Stroke (NINDS 2006a), **meningitis** and **encephalitis** are inflammatory diseases of the membranes that surround the brain and spinal cord and are caused by bacterial or viral infections. Viral meningitis is sometimes called aseptic meningitis to indicate it is not the result of bacterial infection and cannot be treated with antibiotics. Symptoms of encephalitis include sudden fever, headache, vomiting, heightened sensitivity to light, stiff neck and back, confusion and impaired judgment, drowsiness, weak muscles, a clumsy and unsteady gait, and irritability. Symptoms that might require emergency treatment include loss of consciousness, seizures, muscle weakness, or sudden severe dementia.

Symptoms of meningitis, which may appear suddenly, often include high fever, severe and persistent headache, stiff neck, nausea, and vomiting. Changes in behavior such as confusion, sleepiness, and difficulty waking up may also occur. In infants, symptoms of meningitis may include irritability or fatigue, lack of appetite, and fever. Viral meningitis usually resolves in 10 days or less, but other types of meningitis can be deadly if no treated promptly. Anyone experiencing symptoms of meningitis or encephalitis should see a doctor immediately.

ALZHEIMER'S DISEASE

Alzheimer's disease (AD) is a progressive, neurodegenerative disease characterized in the brain by abnormal clumps (amyloid plaques) and tangled bundles of fibers (neurofibrillary tangles) composed of misplaced proteins. Age is the most important risk factor for AD; the number of people with the disease doubles every 5 years beyond age 65 (NINDS 2006b). Three genes have been discovered that cause early onset (familial) AD. Other genetic mutations that cause excessive accumulation of amyloid protein are associated with age-related (sporadic) AD. Symptoms of AD include memory loss, language deterioration, impaired ability to mentally manipulate visual information, poor judgment, confusion, restlessness, and mood swings. Eventually AD destroys cognition, personality, and the ability to function. The early symptoms of AD, which include forgetfulness and loss of concentration, are often missed because they resemble natural signs of aging.

PARKINSON'S DISEASE

Parkinson's disease (PD) belongs to a group of conditions called motor system disorders, which are the result of the loss of dopamine-producing brain cells (NINDS 2006c). The 4 primary symptoms of PD are tremor, or trembling in hand, arms, legs, jaw, and face; rigidity, or stiffness of the limbs and trunk; bradykinesia, or slowness of movement; and postural instability, or impaired balance and coordination. As these symptoms become more pronounced, patients may have difficulty walking, talking, or completing other simple tasks. PD usually affects people over the age of

50. Early symptoms of PD are subtle and occur gradually. In some people the disease progresses more quickly than in others. As the disease progresses, the shaking, or tremor, which affects the majority of PD patients may begin to interfere with daily activities. Other symptoms may include depression and other emotional changes; difficulty in swallowing, chewing, and speaking; urinary problems or constipation; skin problems; and sleep disruptions. There are currently no blood or laboratory tests that have been proven to help in diagnosing sporadic PD, therefore the diagnosis is based on medical history and a neurological examination. The disease can be difficult to diagnose accurately. Doctors may sometimes request brain scans or lab tests in order to rule out other diseases.

Chapter Review Questions

1. The brain and the spinal cord comprise the _____.
2. All nerves of the body residing outside of the brain and spinal cord comprise the
   _____.
3. Sensory neurons are also referred to as _____ neurons while _____ neurons carry motor impulses.
4. The most common neuron is the _____ which communicates from one neuron to another.
5. The branch of the autonomic nervous system that induces the "fight of flight" response is the _____.
6. The cell body of the neuron is the _____.
7. Long extensions off neuronal cell bodies that conduct impulses away from the cell are _____.
8. The dark granular substances inside neuronal cell bodies are called _____.
9. _____ are cells that electrically insulate neuronal axons in the central nervous system.
10. _____ is a CNS disease where the _____ of motor neurons is degenerating or being destroyed.
11. _____ are specialized epithelial cells in the CNS that produce _____.

References and Additional Reading

Brodal, P. 2003. *The Central Nervous System: Structure and Function.* Oxford University Press.
NINDS. 2006a. *Meningitis and Encephalitis Information Page.* Bethesda, MD: National Institute of Neurological Disorders and Stroke.

NINDS. 2006b. *Alzheimer's Disease Information Page.* Bethesda, MD: National Institute of Neurological Disorders and Stroke.

NINDS. 2006c. *Parkinson's Disease Information Page.* Bethesda, MD: National Institute of Neurological Disorders and Stroke.

Sparknotes, 2006. *Structure and Function of Animals.* Accessed August 9, 2006, at www.sparknotes.com.

Taussig, M.1991. *The Nervous System.* New York: Routledge.

CHAPTER 13

Endocrine System

The endocrine system works in concert with the nervous system to control and coordinate the functions of the other organ systems.

Topics in This Chapter

- Types of Glands
- Characteristics of Hormones
- Major Endocrine Glands
- Endocrine System Disorders

The **endocrine system** involves the secretion and release of chemical messengers known as hormones throughout the body via the bloodstream. The organs that make up the endocrine system are called the endocrine glands. Hormones contribute to the control of organ function, growth and development, reproduction and sexual characteristics, how the body uses and stores energy (metabolism), and the volume of fluids and levels of salt and sugar in the blood. The endocrine system works in parallel with the nervous system to control growth and maturation along with homeostasis.

✓   **Important Point:** Chemical messengers from the endocrine system help regulate body activities. Their effect is of longer duration and is more generalized than that of the nervous system.

Note: Much of this chapter is based on information taken from SEER's Training Web site, 2000.

DEFINITION OF KEY TERMS

Before beginning a discussion of the human endocrine system, it is important to list and define pertinent terms:

- **Adrenal**—secretes androgens (male hormones) and aldosterone, which helps maintain the body's salt and potassium balances. Also epinephrine (adrenaline) and norepinepherin (noradrenaline), which are involved in "fight or flight' responses.
- **Antagonistic hormones**—hormones that have an opposite effect on the body.
- **Endocrine glands**—ductless glands that are located throughout the body.
- **Exocrine glands**—secrete nonhormonal chemicals into ducts, which transport the chemicals to a specific location inside and outside the body.
- **Feedback mechanism**—the last step in a series of events controls the first step.
- **Glands**—group of cells that produce and secretes, or gives off, chemicals.
- **Homeostasis**—stable internal environment.
- **Hormone**—is a chemical signal, made in one place and delivered to another that regulates the body's activities.
- **Hypothalamus**—secretes "supervisory" hormone to regulate the pituitary gland.
- **Negative feedback**—release of an initial hormone stimulates release or production of other hormones or substances that subsequently inhibits further release of the initial hormone.
- **Ovaries**—secrete "female" hormones estrogen, which develops and maintains female characteristics, and progesterone, which prepares the uterus for pregnancy.
- **Pancreas**—secretes insulin, the hormones controlling the use of sugar in the body, and other hormones involved with sugar metabolism.
- **Parathyroid**—regulates the use of calcium and phosphorus.
- **Pineal**—secretes melatonin, a hormone involved with daily biological rhythms.
- **Pituitary**—the master gland, secretes hormones that influence many other glands and organs, affecting growth and reproduction.
- **Positive feedback**—release of an initial hormone stimulates release or production of other hormones or substances, which stimulates further release of the initial hormone.
- **Prostaglandins**—group of hormone-like lipids that also regulate cell activities.
- **Receptors**—proteins that are located both inside the cytoplasm and on the surface of a target cell.
- **Steroid hormones**—body synthesizes from cholesterol.
- **Thymus**—stimulates T-cell development of the immune system.
- **Thyroid**—regulates metabolism and blood calcium levels.

✓   **Important Point**: Neurons are the nerve cells that transmit impulses. Supporting cells are called neuroglia.

Types of Glands

There are 2 major categories of glands in the body—exocrine and endocrine:

- **Exocrine glands**—(Greek: *exo*, meaning "outside, beyond," and *krine*, meaning "to separate or secrete") have ducts that carry their secretory product to a surface. These glands include the sweat, sebaceous, and mammary glands, and the glands that secrete digestive enzymes.
- **Endocrine glands**—(Greek: *endo*, meaning "within") consists of a group of organs, often referred to as glands of internal secretion. The endocrine glands do not have ducts to carry their product to a surface. The secretory products of endocrine glands are called hormones and are secreted directly into the blood and then carried throughout the body where they influence only those cells that have receptor sites for that hormone.

✓   **Important Point**: Endocrine glands secrete hormones directly into the blood, which transports the hormones through the body.

Characteristics of Hormones

Chemically, hormones may be classified as either proteins or steroids (i.e., lipids derived form cholesterol). All of the hormones in the human body, except the sex hormones and those from the adrenal cortex, are proteins or protein derivatives.

The so-called *action hormones* are carried by the blood throughout the entire body, yet they affect only certain cells. The specific cells that respond to a given hormone have receptor sites for that hormone. This is sort of a lock and key mechanism. If the key fits the lock, then the door will open. If a hormone fits the receptor site, then there will be an effect. If a hormone and a receptor site do not match, then there is no reaction. All the cells that have receptor sites for a given hormone make up the *target tissue* for that hormone. In some cases, the target tissue is localized in a single gland or organ. In other case, the target tissue is diffuse and scattered throughout the body so that many areas are affected. Hormones bring about their characteristic effects on target cells by modifying cellular activity.

✓   **Important Point**: Cells in a target tissue have receptor sites for specific hormones.

Protein hormones react with receptors on the surface of the cell, and the sequence of events that results in hormone action is relatively rapid. Steroid hormones typically react with receptor sites inside a cell. Because this method of action actually involves synthesis of proteins, it is relatively slow.

Hormones are very potent substances, which means that very small amounts of a hormone may have profound effects on metabolic processes. Because of their potency, hormone secretion must be regulated within very narrow limits in orders to maintain homeostasis in the body.

Many hormones are controlled by some form of a *negative feedback mechanism*. In this type of system, a gland is sensitive to the concentration of a substance that it regulates. A negative feedback system causes a reversal of increases and decreases in body conditions in order to maintain a state of stability or homeostasis. Some endocrine glands secrete hormones in response to other hormones. The hormones that cause secretion of other hormones are called *tropic hormones*. A hormone from gland A causes gland B to secrete its hormone. A third method of regulating hormone secretion is by direct nervous stimulation. A nerve stimulus causes gland A to secrete its hormone.

✓   **Important Point**: Many hormones are regulated by a negative feedback mechanism; some are controlled by other hormones and others are affected by direct nerve stimulation.

Major Endocrine Glands

As mentioned, the endocrine system is made up of the endocrine glands that secrete hormones. Although there are 8 major endocrine glands scattered throughout the body, they are still considered to be a single system because they have similar functions, similar mechanisms of influence, and many important interrelationships.

Although not listed below as 1 of the 8 major endocrine glands, the **hypothalamus** is a collection of specialized cells located in the lower central part of the brain and is the main link between the endocrine and nervous system. Nerve cells in the hypothalamus control the pituitary gland by producing chemicals that either stimulate or suppress hormone secretions from the pituitary.

Some glands have nonendocrine regions that have functions other than hormone secretion. For example, the pancreas has a major exocrine portion that secretes digestive enzymes and an endocrine portion that secretes hormones. The ovaries and testes secrete hormones and also produce the ova and sperm. Some organs, such as the stomach, intestines, and heart, produce hormones, but their primary function is not hormone secretion.

The major glands that make up the human endocrine include the:

- Pituitary
- Pineal
- Thyroid
- Parathyroid
- Adrenal
- Pancreas
- Gonads
- Other Endocrine Glands

PITUITARY GLAND

The pituitary gland or hypophysis (often called the "master gland") is a small gland about 1 centimeter in diameter or the size of a pea. It is nearly surrounded by bone as it rests in the *sella turcica*, a depression in the sphenoid bone. The gland is connected to the hypothalamus of the brain by a slender stalk called the infundibulum. The production and secretion of pituitary hormones can be influenced by factors such as emotions and changes in the seasons. There are 2 distinct regions in the gland: the anterior lobe and the posterior lobe. The activity of the anterior lobe is controlled by releasing hormones from the hypothalamus. The anterior lobe is controlled by nerve stimulation.

Hormones of the Anterior Lobe

1. **Growth hormone** is a protein that stimulates the growth of bones, muscles, and other organs by promoting protein synthesis and the handling of nutrients and minerals. This hormone drastically affects the appearance of an individual because it influences height. If there is too little growth hormone in a child, that person may become a pituitary dwarf of normal proportions but small stature. An excess of the hormone in a child results in an exaggerated bone growth, and the individual becomes exceptionally tall or a giant.
2. **Thyroid-stimulating hormone**, or **thyrotropin**, stimulates the glandular cells of the thyroid to secrete thyroid hormone. When there is a hypersecretion of thyroid-stimulating hormone, the thyroid gland enlarges and secretes too much thyroid hormone.
3. **Adrenocorticotropic hormone** reacts with receptor sites in the cortex of the adrenal gland to stimulate the secretion of cortical hormones, particularly cortisol.
4. **Gonadotropic hormones** react with receptor sites in the gonads, or ovaries and testes, to regulate the development, growth, and function of these organs.
5. **Prolactin hormone** promotes the development of glandular tissue in the female breast during pregnancy and stimulates (activates) milk production after the birth of the infant.

Hormones of the Posterior Lobe

1. **Antidiuretic hormone** promotes the reabsorption of water by the kidney tubules, with the result that less water is lost as urine. This mechanism conserves water for the body. Insufficient amounts of antidiuretic hormone cause excessive water loss in the urine.
2. **Oxytocin** causes contraction of the smooth muscle in the wall of the uterus. It also stimulates the ejection milk from the lactating breast.

PINEAL GLAND

The **pineal gland**, also called pineal body, is a small cone-shaped structure (in the middle of the brain) that extends posteriorly from the third ventricle of the brain. The pineal gland consists of portions of neurons, neuroglial cells, and specialized secretory cells called pinealocytes. The pinealocytes synthesize the hormone melatonin and secrete it directly into the cerebrospinal fluid, which takes it into the blood. Melatonin affects reproductive development and daily physiologic cycles (it may help regulate sleep at night and waking in the morning).

THYROID GLAND

The **thyroid gland**, located in the front part of the lower neck, consists of 2 lobes, 1 on each side of the trachea, just below the larynx or voice box; it is shaped like a bow tie or butterfly. The 2 lobes are connected by a narrow band of tissue called the isthmus. Internally, the gland consists of follicles, which produce thyroxine and triiodothyronine hormones. These hormones contain iodine and control the rate at which cells burn fuels from food to produce energy.

About 95% of the active thyroid hormone is thyroxine, and most of the remaining 5% is triiodothyronine. Both of these require iodine for their synthesis. Thyroid hormone secretion is regulated by a negative feedback mechanism that involves the amount of circulating hormone, hypothalamus, and adenohypophysis.

If there is an iodine deficiency, the thyroid cannot make sufficient hormone. This stimulates the anterior pituitary to secrete thyroid-stimulating hormone, which causes the thyroid gland to increase in size in a vain attempt to produce more hormones. But it cannot produce more hormones because it does not have the necessary raw material, iodine. This type of thyroid enlargement is called *simple goiter* or *iodine deficiency goiter*.

Calcitonin is secreted by the parafollicular cells of the thyroid gland. This hormone opposes the action of the parathyroid glands by reducing the calcium level in the blood. If blood calcium becomes too high, calcitonin is secreted until calcium ion levels decrease to normal.

PARATHYROID GLAND

Attached to the thyroid are 4 tiny glands (embedded in the connective tissue capsule on the posterior surface of the thyroid) that function together, called the **parathyroids**. They release parathyroid hormone, which regulates the level of calcium in the blood with the help of calcitonin, which, as mentioned, is produced in the thyroid.

Hypoparathyroidism, or insufficient secretion of parathyroid hormone, leads to increased nerve excitability. The low blood calcium levels trigger spontaneous and continuous nerve impulses, which then stimulate muscle contraction.

ADRENAL GLAND

The **adrenal**, or **suprarenal, gland** is paired with 1 gland located near the upper portion of each kidney. Each gland is divided into an outer cortex and an inner medulla.

The cortex and medulla of the adrenal gland, like the anterior and posterior lobs of the pituitary, develop from different embryonic tissues and secrete different hormones. The adrenal cortex is essential to life, but the medulla may be removed with no life-threatening effects.

The hypothalamus influences both portions of the adrenal gland but by different mechanisms. The adrenal cortex is regulated by negative feedback involving the hypothalamus and adrenocorticotropic hormone; the medulla is regulated by nerve impulses from the hypothalamus.

The **adrenal cortex** consists of 3 different regions, with each region producing a different group of hormones. Mineralocorticoids are secreted by the outermost region of the adrenal cortex. The principal mineralocorticoid is aldosterone, which acts to conserve sodium ions and water in the body. Glucocorticoids are secreted by the middle region of the adrenal cortex. The principal glucocorticoid is cortisol. The third group of steroids secreted by the adrenal cortex is the gonadocorticoids, or sex hormones. These are secreted by the innermost region. Male hormones (androgens) and female hormones (estrogens) are secreted in minimal amounts in both sexes by the adrenal cortex, but their effect is usually masked by the hormones from the testes and ovaries. In females, the masculinization effect of androgen secretion may become evident after menopause, when estrogen levels from the ovaries decrease.

✓   **Important Point**: Chemically, all the cortical hormones are steroid.

The **adrenal medulla** develops from neural tissue and secretes 2 hormones, epinephrine (also called adrenaline) and norepinephrine. These 2 hormones are secreted in response to stimulation by sympathetic nerve, particularly during stressful situations. A lack of hormones from the adrenal medulla produces no significant effects. Hypersecretion, usually from a tumor, causes prolonged or continual sympathetic responses.

PANCREAS

The **pancreas** is also part of the body's hormone-secreting system, even though it is also associated with the digestive system because it produces and secretes digestive enzymes. The pancreas is a long, soft organ that lies transversely along the posterior abdominal wall, posterior to the stomach, and extends from the region of the duodenum to the spleen. In addition to others, the pancreas produces 2 important hormones, **insulin** and **glucagon**. They work together to maintain a steady level of glucose, or sugar, in the blood and to keep the body supplied with fuel to produce and maintain stores of energy.

GONADS

The **gonads**, the primary reproductive organs, are the testes in the male and the ovaries in the female. These organs are responsible for producing the sperm and ova, but they also secrete hormones and are considered to be endocrine glands.

Testes

Male sex hormones, as a group, are called androgens. The principal androgen is testosterone, which is secreted by the testes. A small amount is also produced by the adrenal cortex. Production of testosterone begins during fetal development, continues for a short time after birth, nearly ceases during childhood, and then resumes at puberty. This steroid hormone is responsible for:

- The growth and development of the male reproductive structures
- Increased skeletal and muscular growth
- Enlargement of the larynx accompanied by voice changes
- Growth and distribution of body hair
- Increased male sexual drive

Testosterone secretion is regulated by a negative feedback system that involves releasing hormones from the hypothalamus and gonadotropins form the anterior pituitary.

Ovaries

Two groups of female sex hormones are produced in the ovaries: the estrogens and progesterone. These steroid hormones contribute to the development and function of the female reproductive organs and sex characteristics. At the onset of puberty, estrogens promote:

- The development of the breasts
- Distribution of fat evidenced in the hips, legs, and breast
- Maturation of reproductive organs such as the uterus and vagina

Progesterone causes the uterine lining to thicken in preparation for pregnancy. Together, progesterone and estrogens are responsible for the changes that occur in the uterus during the female menstrual cycle.

OTHER ENDOCRINE GLANDS

In addition to the major endocrine glands, other organs have some hormonal activity as part of their function. These include the thymus, stomach, small intestines, heart, and placenta.

- **Thymosin**, produced by the thymus gland, plays an important role in the development of the body's immune system.
- The lining of the stomach, **the gastric mucosa**, produces a hormone called *gastrin* in response to the presence of food in the stomach. This hormone stimulates the

production of hydrochloric acid and the enzyme pepsin, which are used in the digestion of food.

- **The mucosa of the small intestine** secretes the hormones secretin and cholecystokinin. Secreting stimulates the pancreas to produce a bicarbonate-rich fluid that neutralizes the stomach acid. Cholecystokinin stimulates contraction of the gallbladder, which releases bile. It also stimulates the pancreas to secrete digestive enzyme.
- The **heart** also acts as an endocrine organ in addition to its major role of pumping blood. Special cells in the wall of the upper chambers of the heart, called atria, produce a hormone called atrial natriiuretic hormone, or atriopeptin.
- The **placenta** develops in the pregnant female as a source of nourishment and gas exchange for the developing fetus. It also serves as a temporary endocrine gland. One of the hormones it secretes is human chorionic gonadotropins, which signals the mother's ovaries to secrete hormones to maintain the uterine lining so that it does not degenerate and slough off in menstruation.

Endocrine System Disorders

Too much or too little of any hormone can be harmful to the body. For example, if the pituitary gland produces too much growth hormone, a teen may grow excessively tall. If it produces too little, a teen may be unusually short. Doctors can often treat problems with the endocrine system by controlling the production of hormones or replacing certain hormones with medication. Some endocrine problems that affect people include:

- **Addison's disease** (also called **adrenal insufficiency**)—This condition is an endocrine or hormonal disorder that occurs in all age groups and afflicts men and women equally. The disease is characterized by weight loss, muscle weakness, fatigue, low blood pressure, and sometimes darkening of the skin in both exposed and nonexposed parts of the body. Addison's disease occurs when the adrenal glands do not produce enough of the hormone cortisol and, in some case, the hormone aldosterone. Doctors treat Addison's disease with medications to replace corticosteroid hormones.
- **Diabetes**—This condition is marked by high levels of glucose resulting from defects in insulin production, insulin action, or both. Diabetes can lead to serious complications and premature death, but people with diabetes can take steps to control the disease and lower the risk of complications. Approximately 21 million Americans have diabetes—7% of the U.S. population. Of these, 6.2 million do not know they have the disease. Each year, about 1.5 million people age 20 or older are diagnosed with diabetes. The number of people diagnosed with diabetes has risen from 1.5 million in 1958 to 14.6 million in 2005, an increase of epidemic proportions (NIH 2005).

  - **Type 1 diabetes**—When the pancreas fails to produce enough insulin, type 1 diabetes occurs. Type 1 accounts for 5%–10% of all diagnosed cases of diabetes.
  - **Type 2 diabetes**—Unlike type 1 diabetes, in which the body can't produce normal amounts of insulin, in type 2 diabetes the body can't respond to insulin normally. Type 2 diabetes accounts for 90%–95% of all diagnosed cases of diabetes.

- **Pancreatitis**—Pancreatitis is an inflammation of the pancreas. Normally, digestive enzymes do not become active until they reach the small intestine, where they begin digesting food. But if these enzymes become active inside the pancreas, they start "digesting" the pancreas itself.

  - **Acute pancreatitis** can be a severe, life-threatening illness with many complications. About 80,000 cases occur in the U.S. each year; some 20% of them are severe. Acute pancreatitis usually begins with pain in the upper abdomen that may last for a few days. The pain may be severe and may become constant—just in the abdomen—or it may reach to the back and other areas. It may be sudden and intense or begin as a mild pain that gets worse when food is eaten. Someone with acute pancreatitis often looks and feels very sick. Acute pancreatitis occurs more often in men and women. Acute pancreatitis is usually caused by gallstones or by drinking too much alcohol, but these aren't the only causes. If alcohol use and gallstones are ruled out, other possible causes of pancreatitis should be carefully examined so that appropriate treatment—if available—can begin (NIH 2004).

  - **Chronic pancreatitis** occurs when digestive enzymes attack and destroy the pancreas and nearby tissues, causing scarring and pain. The usual cause of chronic pancreatitis is many years of alcohol abuse but the chronic form may also be triggered by only one acute attack, especially if the pancreatic ducts are damaged. The damaged ducts cause the pancreas to become inflamed, tissue to be destroyed, and scar tissue to develop (NIH 2004).

  ✓   **Important Point**: While common, alcoholism is not the only cause of chronic pancreatitis.

- **Hyperparathyroidism**—If the parathyroid glands secrete too much hormone, as happens in primary hyperparathyroidism, the balance is disrupted: blood calcium rises. This condition of excessive calcium in the blood, called hypercalcemia, is what usually signals the doctor that something may be wrong with the parathyroid glands. In 85% of people with primary hyperparathyroidism, a benign tumor called an adenoma has formed on one of the parathyroid glands, causing it to become overactive. Benign tumors are noncancerous. In most other cases, the excess hormone comes from 2 or more enlarged parathyroid glands, a condition called hyperplasia. This excess parathyroid hormone (PTH) triggers the release of too much calcium into the bloodstream. The bones lose calcium, and too much calcium may be absorbed from food. The levels of calcium may increase in the urine, causing kidney stones. PTH also lowers blood phosphorous levels by increasing excretion of phosphorus in the urine. Doctors often treat this problem with medication. Very rarely, hyperparathyroidism is caused by cancer of a parathyroid gland (NIH 2006).

- **Thyroid disease**—The thyroid secretes the hormone thyroxine which speeds up metabolism and helps manage growth and development. When the thyroid produces too much hormone, the body uses energy faster than it should. This condition is called hyperthyroidism. When the thyroid doesn't produce enough hormones, the body uses energy slower than it should. This condition is called hypothyroidism. There are many different reasons why either of these conditions might develop. However, women are 5 to 8 times more likely than men to have thyroid problems.

Chapter Review Questions

1. The _____ system involves the secretion and release of hormones.
2. The _____ helps maintain the body's salt and potassium balances.
3. Stable internal environment: _____.
4. Secretes melatonin: _____.
5. _____ glands have ducts that carry their secretory products to the surface.
6. Hormones that cause secretion of other hormones are called _____ hormones.
7. _____ causes contraction of the smooth muscle in the wall of the uterus.
8. Ninety-five percent of the active thyroid hormone is _____.
9. Secretes epinephrine: _____.
10. The pancreas produces insulin and _____.
11. The _____ are the primary reproductive organs.
12. _____ diabetes occurs when the body (pancreas) does not produce normal amounts of insulin.

References and Additional Reading

Bilezekian, J.P., R. Marcus, and M. Levine, eds. *The Parathyroids*, 2nd ed. New York: Academic Press.

Bilezikian J.P., et al. 2002. Summary Statement from a Workshop on Asymptomatic Primary Hyperparathyroidism: A Perspective for the 21st Century. *Journal of Bone Mineral Research* 17(Supp 2): N2–N11.

Chrousos, G.P. 2001. Glucocorticoid Therapy. In *Endocrinology and Metabolism*, eds. P. Felig and L. Frohman, 4th ed. New York: McGraw-Hill.

Favus, M.J., et al., eds. 2003. *Primer on the Metabolic Bone Diseases and Disorders of Mineral Metabolism,* 5th ed. Philadelphia: Lippincott Raven. American Society of Bone and Mineral Research.

Marx, S.J. 2000. Hyperparathyroid and Hypoparathyroid Disorders. *New England Journal of Medicine* 343: 1863–1875.

Miller, W., and G.P. Chrousos. 2001. The Adrenal Cortex. In *Endocrinology and Metabolism*, eds. P. Felig and L. Frohman, 4th ed. New York: McGraw-Hill.

NIH. 2004. *Pancreatitis*. Bethesda, MD: National Institutes of Health.

NIH. 2005. *National Diabetes Statistics Fact Sheet: General Information and National Estimates on Diabetes in the United States*. Bethesda, MD: U.S. Department of Health and Human Services.

NIH. 2006. *Hyperparathyroidism*. Bethesda, MD: U.S. Department of Health and Human Services.

SEER. 2000. *Hormones*. Accessed August 19, 2006, at www.training.seer.cancer.gov.

Stewart, P.M. 2003. The Adrenal Cortex. In *Williams Textbook of Endocrinology*, ed. P. Larsen, 10th ed. Philadelphia: Saunders.

Ten, S., M. New, and N. Maclaren. 2001. Clinical Review 130: Addison's Disease. *Journal of Clinical Endocrinology & Metabolism* 86:2909–2922.

CHAPTER 14

Reproductive System

As dictated by evolution, an organism's purpose is to reproduce and ensure the survival of the species.

Topics in This Chapter

- Male Reproductive System
- Female Reproductive System
- Reproductive System Disorders
- Human Development

The ability to reproduce is one of the unifying characteristics of all living things. The major function of the reproductive system is to ensure survival of the species. Other systems in the body, such as the endocrine and urinary systems, work continuously to maintain homeostasis for survival of the individual. An individual may live a long, healthy, and happy life without producing offspring, but if the species is to continue, at least some individuals must produce offspring. Production of offspring is accomplished via sexual reproduction, which produces offspring that are genetically different from their parents.

✓ **Important Point**: Within the context of producing offspring, the reproductive system has 4 functions:

- To produce egg and sperm cells
- To transport and sustain these cells
- To nurture the developing offspring
- To produce hormones

Note: Much of the material in this chapter is based on material from SEER's (2000) *Reproduction* training module, accessible at www.training.seer.cancer.gov/module_anatomy.

Sexual reproduction occurs when 2 haploid gametes, 1 from each parent, fuse to form a zygote which develops into an offspring genetically different from the parents. This fertilization can take place externally, as is the case for many aquatic organisms that release their unfertilized gametes into the water, or internally. As animals have evolved, they have developed special structures for the production of gametes, for the fertilization process, and for the support and nourishment of the developing young. Collectively, these structures are known as the reproductive system. The anatomy of the male and the female reproductive system is significantly different.

Before beginning a discussion of the male and female reproductive systems, it is important to define terms pertinent to this discussion.

DEFINITION OF KEY TERMS

- **Amniotic sac**—a bubble-like sac that surrounds the baby while it is inside of the mother's uterus. The amniotic sac is filled with amniotic fluid.
- **Cervix**—the narrow end of the uterus which has a small opening that connects the uterus with the vagina.
- **Conception**—the beginning of pregnancy, when the sperm cell from the father joins with the egg cell from the mother.
- **Contractions**—the squeezing action of the uterine muscle that helps to push the baby out of the mother during the process of birth.
- **Embryo**—the developing baby during the first 2 months of pregnancy.
- **Ejaculation**—the process by which semen is ejected from the penis.
- **Epididymis**—a storage chamber in the male's body that's attached to each testicle. This is where sperm cells are nourished and mature.
- **Erection**—the penis becomes stiff and hard due to increased blood flow. Erections may happen in response to physical or emotional stimulation, or sometimes an erection happens for no reason at all.
- **Estrogen**—the main female sex hormone produced by the ovaries.
- **Fallopian tubes**—narrow tubes that are connected to the uterus. The fringes of the Fallopian tube catch the egg cell when it is released from the ovary and then the egg cell slowly travels from the ovary to the uterus.
- **Fertilization**—this is what happens when a male sperm cell unites with a female egg cell. A fertilized egg cell grows into a baby.
- **Fetus**—the developing baby from 2 months of pregnancy until birth.
- **Genitals**—the external sex organs.
- **Hormones**—chemical messengers created by glands that control specific things that happen in the body.
- **Menstruation**—the process by which the lining of the uterus is shed periodically as menstrual flow. It usually happens about once a month except during pregnancy.
- **Nocturnal emission**—an ejaculation of semen that happens while a male is sleeping.
- **Ovaries**—two small organs inside of a female's body where the egg cells are produced and stored. Each ovary is about the size of a walnut, and there is one on each side of the uterus. The ovaries also produce the hormones estrogen and progesterone.

- **Ovulation**—the release of an egg cell by an ovary. This process usually occurs at the midpoint of the menstrual cycle.
- **Ovum**—another name for the female egg cell. It is smaller than a grain of salt.
- **Penis**—the male reproductive organ involved in sexual intercourse and elimination of urine.
- **Period**—the days when menstruation is taking place.
- **Placenta**—a structure that grows from the uterine wall that allows nutrients and oxygen from the mother to pass through the umbilical cord to the baby and waste products from the baby go back through the placenta.
- **Progesterone**—a female hormone.
- **Prostrate gland**—a male gland at the base of the bladder. It contributes a thin, milky fluid that makes up the largest part of the semen.
- **Puberty**—the stage of growth where a child's body turns into the body of an adult.
- **Scrotum**—the outside sac of loose skin under the penis that holds the testicles.
- **Semen**—a milky white fluid made by the seminal vesicles and prostate gland. This fluid mixes with the sperm cells during an ejaculation. About a teaspoon of semen comes out of the penis during ejaculation.
- **Seminal vesicle**—one of 2 glands located behind the male bladder, which secrete a fluid that forms part of semen.
- **Sexual intercourse**—the erect penis of the male entering the vagina of the female.
- **Sperm**—the tiny microscopic cells produced by a male that contain the genes from the father. A sperm cell from the father must join with an egg cell from the mother for a baby to be created.
- **Testicles**—the main male reproductive glands in which sperm are produced. The testicles also produce the main male hormone testosterone.
- **Testosterone**—a male sex hormone, which causes a boy's body to develop into a man's. Testosterone is responsible for the development of more muscle mass, deeper voice, and facial hair.
- **Umbilical cord**—a cord attaching the unborn child to the mother. This cord comes from the placenta inside of the uterus and attaches to the baby's abdomen. The umbilical cord is the lifeline between mother and baby.
- **Uterus**—a muscular organ about the size and shape of an upside-down pear, located inside the body between a female's hips. Also called the womb. This is where the baby grows before it is born.
- **Vagina**—the tube-shaped passageway that leads from the uterus to the outside of the body.
- **Vas deferens**—a thin tube that transports sperm from the epididymis (the storage chamber attached to the back of each testicle) to the urethra.
- **Vulva**—the entire outside genital area of a female.
- **Womb**—another name for the uterus.

✓ **Important Point**: The primary reproductive organs are the gonads, which produce the gametes and hormones. The secondary, or accessory, structures transport and sustain the gametes and nurture the developing offspring.

Male Reproductive System

The male reproductive system has 2 major functions: (1) it produces **sperm** cells, the male gametes, through the process of *spermatogenesis*; and (2) it produces semen, a fluid that acts as a vehicle and nourishment for sperm as they make their way through the female reproductive system on their way to fertilize the egg.

The male reproductive system, like that of the female, consists of those organs whose function is to produce a new individual, i.e., to accomplish reproduction. This system consists of a pair of testes and a network of excretory ducts including the epididymis, ductus deferens (vas deferens), and ejaculatory ducts; seminal vesicles; the prostrate and bulbourethral glands; and the penis.

TESTES

The **testes** are the male gonads: they produce the sperm, which, as mentioned, is the male gamete. The male gonads, testes, or testicles, begin their development high in the abdominal cavity, near the kidneys. During the last 2 months before birth, or shortly after birth, they descend through the inguinal canal into the scrotum, a pouch that extends below the abdomen, posterior to the penis. Although this location of the testes, outside the abdominal cavity, may seem to make them vulnerable to injury, it provides a temperature about 3°C below normal body temperature. This lower temperature is necessary for the production of viable sperm. The scrotum consists of skin and subcutaneous tissue. A vertical septum, or partition, of subcutaneous tissue in the center divides it into 2 parts, each containing 1 testis. Smooth muscle fibers (called "the dartos muscle") in the subcutaneous tissue contract to give the scrotum its wrinkled appearance. When these fibers are relaxed, the scrotum is smooth. Another muscle, the cremaster muscle, consists of skeletal muscle fibers and controls the position of the scrotum and testes. When it is cold or a man is sexually aroused, this muscle contracts to pull the testes closer to the boy for warmth.

✓   **Interesting Point**: Each testis is an oval structure about 5 cm long and 3 cm in diameter. The testes' location within the scrotum is necessary for the production of viable sperm.

Sperm are produced by **spermatogenesis** within the seminiferous tubules. A transverse section of a seminiferous tubule shows that it is packed with cells in various stages of development. Interspersed with these cells, there are large cells that extend from the periphery of the tubule to the lumen. These large cells are the supporting, or sustentacular, cells (Sertoli's cells), which support and nourish the other cells.

Spermatogenesis (and oogenesis in the female) differs from mitosis because the resulting cells have only half the number of chromosomes as the original cell. When the sperm cell nucleus unites with an egg cell nucleus, the full number of chromosomes is restored. If sperm and egg cells were produced by mitosis, then each successive generation would have twice the number of chromosomes as the preceding one.

The final step in the development of sperm is called **spermiogenesis**. In this process, the spermatids formed from spermatogenesis become mature spermatozoa, or sperm. The mature sperm cell has a head, midpiece, and tail. The head, also called the nuclear region, contains the 23 chromosomes surrounded by a nuclear membrane. The tip of the head is covered by an acrosome, which contains enzymes that help the sperm penetrate the female gamete. The midpiece, metabolic region, contains mitochondria that provide adenosine triphosphate (ATP). The tail, locomotor region, uses a typical flagellum for locomotion. The sperm are released into the lumen of the seminiferous tubule and leave the testes. They then enter the epididymis where they undergo their final maturation and become capable of fertilizing a female gamete.

✓   **Important Point**: Sperm production begins at puberty and continues throughout the life of a male. The entire process, beginning with a primary spermatocyte, takes about 74 days. After ejaculation, the sperm can live for about 48 hours in the female reproductive tract.

DUCT SYSTEM

Sperm cells pass through a series of ducts to reach the outside of the body. After they leave the testes, the sperm passes through the epididymis, ductus deferens, ejaculatory duct, and urethra.

Sperm leave the testes through a series of efferent ducts that enter the **epididymis**. Each epididymis is a long tube (about 6 meters) that is tightly coiled to form a comma-shaped organ located along the superior and posterior margins of the testes. When the sperm leave the testes, they are immature and incapable of fertilizing ova. They complete their maturation process and become fertile as they move through the epididymis. Mature sperm are stored in the lower portion, or tail, of the epididymis.

The **ductus deferens**, also called **vas deferens**, is a fibromuscular tube that is continuous (or contiguous) with the epididymis. It begins at the bottom (tail) of the epididymis then turns sharply upward along the posterior margin of the testes. The ductus deferens enters the abdominopelvic cavity through the inguinal canal and passes along the lateral pelvic wall. It crosses over the ureter and posterior portion of the urinary bladder, and then descends along the posterior wall of the bladder toward the prostate gland. Just before it reaches the prostate gland, each ductus deferens enlarges to form an ampulla. Sperm are stored in the proximal portion of the ductus deferens, near the epididymis, and peristaltic movements propel the sperm through the tube.

The proximal portion of the ductus deferens is a compound of the spermatic cord, which contains vascular and neural structures that supply the testes. The spermatic cord contains the ductus deferens, testicular artery and veins, lymph vessels, testicular nerve, cremaster muscle that elevates the testes for warmth and at times of sexual stimulation, and a connective tissue covering.

Each ductless deferens, at the ampulla, joins the duct from the adjacent seminal vesicle (one of the accessory glands) to form a short ejaculatory duct. Each ejaculatory duct passes through the prostate gland and empties into the urethra.

The **urethra** extends from the urinary bladder to the external urethral orifice at the tip of the penis. It is a passageway for sperm and fluids from the reproductive system and urine from the urinary system. While reproductive fluids are passing though the urethra, sphincters contract tightly to keep urine from entering the urethra.

The male urethra is divided into 3 regions. The prostatic urethra is the proximal portion that passes through the prostate gland. It receives the ejaculatory duct, which contains sperm and secretions from the seminal vesicles, and numerous ducts from the prostate glands. The next portion, the membranous urethra, is a short region that passes through the pelvic floor. The longest portion is the penile urethra, which extends the length of the penis and opens to the outside at the external urethral orifice. The ducts from the bulbourethral glands open into the penile urethra.

ACCESSORY GLANDS

The accessory glands of the male reproductive system are the seminal vesicles, prostate gland, and the bulbourethral glands. These glands secrete fluids that enter the urethra.

- **Seminal vesicles**—the paired seminal vesicles are saccular glands posterior to the urinary bladder. Each gland has a short duct that joins with the ductus deferens at the ampulla to form an ejaculatory duct, which then empties into the urethra.
- **Prostate**—is a firm, dense structure that is located just inferior to the urinary bladder. It is about the size of a walnut and encircles the urethra as it leaves the urinary bladder. Numerous short ducts from the substance of the prostate gland empty into the prostate urethra. Secretions of the prostate function to enhance the motility of the sperm.
- **Bulbourethral glands**—the paired bulbourethral (Cowper's) glands are small, about the size of a pea, and located near the base of the penis. A short duct form each gland enters the proximal end of the penile urethra. In response to sexual stimulation, the bulbourethral glands secrete fluid which helps to neutralize the acidity of the vagina, and provides some lubrication for the tip of the penis during intercourse.
- **Seminal fluid**—is a slightly alkaline mixture of sperm cells and secretions from the accessory glands. Secretions from the seminal vesicles make up about 60% of the volume of the semen.

PENIS

The male copulatory organ, the **penis**, is a cylindrical pendant organ located anterior to the scrotum and functions to transfer sperm to the vagina. The penis consists of 3 columns of erectile tissue that are wrapped in connective tissue and covered with skin.

MALE SEXUAL RESPONSE

The male sexual response includes erection and orgasm accompanied by ejaculation of semen. Orgasm is followed by a variable time period during which it is not possible to achieve another erection.

 Important Point: The male reproductive system consists of the testes, duct system, accessory glands, and penis.

Female Reproductive System

While the male reproductive system is designed to produce and deposit sperm in the female, the female reproductive system has the more formidable task of receiving the male gametes, producing the female gametes, and, in the event of fertilization, maintaining and supporting a pregnancy.

"The organs of the female reproductive system produce and sustain the female sex cells (egg cells or ova), transport these cells to a site where they may be fertilized by sperm, provide a favorable environment for the developing fetus, move the fetus to the outside at the end of the development period, and produce the female sex hormones. The female reproductive system includes the ovaries, Fallopian tubes, uterus, vagina, accessory glands, and external genital organs" (The Universe Within Touring Company 2006).

OVARIES

The primary female reproductive female organs, or gonads, are the 2 **ovaries**. Each ovary is a solid, ovoid structure about the size and shape of an almond, about 35 cm in length, 2 cm wide, and 1 cm thick. The ovaries are located in the shallow depressions, called ovarian fossae, one on each side of the uterus, in the arterial walls of the pelvic cavity. They are held loosely in place by peritoneal ligaments.

Female sex cells, or gametes, develop in the ovaries by a form of meiosis called **oogenesis**. The sequence of events in oogenesis is similar to the sequence in spermatogenesis, but the timing and final result are different. Early in fetal development, primitive germ cells in the ovaries differentiate into oogonia. These divide rapidly to form thousands of cells, still called oogonia, which have a full complement of 46 (23 pairs) chromosomes. Oogonia then enter a growth phase, enlarge, and become primary oocytes. The diploid (46 chromosomes) primary oocytes replicate their DNA and begin the first meiotic division, but the process stops in prophase and the cells remain in this suspended state until puberty. In regards to female egg production after puberty, Halperin (2004) notes that "women are thought to be born with a fixed number of eggs that are released between puberty and menopause."

✓   **Important Point**: Ovulation, prompted by luteinizing hormone from the anterior pituitary, occurs when the mature follicle at the surface of the ovary ruptures and releases the secondary oocytes into the peritoneal cavity.

FALLOPIAN TUBES

There are 2 uterine tubes, 1 associated with each ovary, called **Fallopian tubes** or **oviducts**. The end of the tube near the ovary expands to form a funnel-shaped infundibulum, which is surrounded by fingerlike extensions called fimbriae. Because there is no direct connection between the infundibulum and the ovary, the oocyte enters the peritoneal cavity before it enters the Fallopian tube. At the time of ovulation, the fimbriae increase their activity and create currents in the peritoneal fluid that help propel the oocyte into the Fallopian tube. Once inside the Fallopian tube, the oocyte is moved along by the rhythmic beating of cilia on the epithelial lining and by peristaltic action of the smooth muscle in the wall of the tube. The journey through the Fallopian tube takes about 7 days. Because the oocyte is fertile for only 24 to 48 hours, fertilization usually occurs in the Fallopian tube.

UTERUS

The **uterus** is a muscular organ that receives the fertilized oocyte and provides an appropriate environment for the developing fetus. Before the first pregnancy, the uterus is about the size and shape of a pear, with the narrow portion directed inferiorly. After childbirth, the uterus is usually larger, and then regresses after menopause.

VAGINA

The **vagina** is a fibromuscular tube, about 10 cm long that extends from the cervix of the uterus to the outside. It is located between the rectum and the urinary bladder. Because the vagina is tilted posteriorly as it ascends and the cervix is tilted anteriorly, the cervix projects into the vagina at nearly a right angle. The vagina serves as a passageway for menstrual flow, receives the erect penis during intercourse, and is the birth canal during childbirth.

EXTERNAL GENITALIA

The **external genitalia** are the accessory structures of the female reproductive system that are external to the vagina. They are also referred to as the vulva or pudendum. The external genitalia include the labia majora, mons pubis, labia minora, clitoris, and glands within the vestibule.

The clitoris is an erectile organ, similar to the male penis, that responds to sexual stimulation. Posterior to the clitoris, the urethra, vagina, paraurethral glands, and greater vestibular glands open into the vestibule.

✓   **Important Point**: The female reproductive system includes the ovaries, uterine tubes, uterus, vagina, accessory glands, and external genital organs.

FEMALE SEXUAL RESPONSE

The female sexual response includes arousal and orgasm, but there is no ejaculation. A woman may become pregnant without having an organism. Various hormones play an important role in female sexual response: follicle-stimulating hormone, luteinizing hormone, estrogen, and progesterone have major roles in regulating the functions of the female reproductive system.

At puberty, when the ovaries and uterus are mature enough to respond to hormonal stimulation, certain stimuli cause the hypothalamus to start secreting gonadotropin-releasing hormone. This hormone enters the blood and goes to the anterior pituitary gland where it stimulates the secretion of follicle-stimulating hormone and luteinizing hormone. These hormones, in turn, affect the ovaries and uterus and the monthly cycle begin. A woman's reproductive cycles last from menarche to menopause.

The monthly ovarian cycle begins with the follicle development during the follicular phase, continues with ovulation during the ovulatory phase, and concludes with the development and regression of the corpus luteum during the luteal phase.

The uterine cycle takes place simultaneously with the ovarian cycle. The uterine cycle begins with menstruation during the menstrual phase, continues with repair of the endometrium during the proliferative phase, and ends with the growth of glands and blood vessels during the secretory phase.

Menopause occurs when a woman's reproductive cycles stop. This period is marked by decreased levels of ovarian hormones and increased levels of pituitary follicle-stimulating hormone and luteninzing hormone. The changing hormone levels are responsible for the symptoms associated with menopause.

MAMMARY GLANDS

The female reproductive system also consists of the **mammary glands**, which produce milk to nourish the young. Functionally, as mentioned, the mammary glands produce milk; structurally, they are modified sweat glands. Mammary glands, which are located in the breasts overlying the pectoralis major muscles, are present in both sexes, but usually are functional only in the female.

Externally, each breast has a raised nipple, which is surrounded by a circular pigmented area called the areola. The nipples are sensitive to touch, due to the fact they contain smooth muscle that contracts and causes them to become erect in response to stimulation.

Internally, the adult female breast contains 15 to 20 lobes of glandular tissue that radiate around the nipple. The lobes are separated by connective tissue and adipose. The connective tissue helps support the breast. Some bands of connective tissue, called suspensory ligaments, extend through the breast from the skin to the underlying muscles. The amount and distribution of the adipose tissue determines the size and shape of the breast. Each lobe consists of lobules that contain the glandular units. A lactiferous duct collects the milk from the lobules within each lobe and carries it to the nipple. Just before the nipple the lactiferous duct enlarges to form a lactiferous sinus (ampulla), which serves as a reservoir for milk. After the sinus, the duct again narrows and each duct opens independently on the surface of the nipple.

Hormones regulate mammary gland function. At puberty, increasing levels of estrogen stimulate the development of glandular tissue in the female breast. Estrogen also causes the breast to increase in size through the accumulation of adipose tissue. Progesterone stimulates the development of the duct system. During pregnancy these hormones enhance further development of the mammary glands. Prolactin from the anterior pituitary stimulates the production of milk within the glandular tissue, and oxytocin causes the ejection of milk from the glands.

✓   **Important Point**: Estrogen and progesterone stimulate the development of glandular tissue and ducts in the breast. Prolactin stimulates the production of milk, and oxytocin causes the ejection of milk.

Reproductive System Disorders

In the male reproductive system, 2 of the most significant disorders include prostrate and testicular cancer. In the female reproduction system, ovarian and uterine cancer is significant.

- **Prostate cancer**—is made up of cells that do not grow normally. The cells divide and create new cells the body does not need, forming a mass of tissue called a tumor. The abnormal cells sometimes spread to other parts of the body, multiply, and cause death. As with many types of cancers, medical experts do not know what causes prostrate cancer. They are studying several possible causes. Moreover, medical experts do not know how to prevent prostate cancer, but they are studying many factors. They do know that not smoking, maintaining a healthy diet, staying physically active, and seeing a doctor regularly contribute to overall good health.
- **Testicular cancer**—type of cancer that develops in the testicles. Testicular cancer has one of the highest cure rates of all cancers: in excess of 90%.
- **Ovarian cancer**—is cancer that begins in the ovaries. Many types of tumors can start growing in the uterus. Women with these types of tumors can be treated successfully by removing one ovary or the part of the ovary that contains the tumor. Other types of ovarian tumors are malignant (cancerous) and can spread to other parts of the body. Their treatment is more complex.

• **Uterine cancer**—is a cancer that starts in the endometrium, the inner lining of the uterus (womb). Nearly all endometrial cancers (about 95%) are cancers of the glandular cells.

Human Development

In biological terms, **human development** is the process of growing to maturity. This entails growth from a one-celled zygote to an adult human being, outlined below. Stages of age-related physical development include:

1. **Zygote**—the point of conception
2. **Blastocyst**—the period between conception and embryonic stages
3. **Embryo stage**—starts at 3 weeks and continues until the end of 8th week of pregnancy
4. **Fetus stage**—begins at the end of 8th week and continues until childbirth
5. **Birth**
6. **Child**
7. **Adolescence/puberty**
8. **Adult**
9. **Death**—occurs at various ages

Chapter Review Questions

1. The _____ system functions for the survival of the species.
2. The primary reproductive organs, also called gonads, are the _____, which are responsible for producing the egg and sperm cells.
3. Spermatogenesis differs from mitosis because the resulting cells have _____ the number of chromosomes as the original cell.
4. The final step in the development of sperm is called _____.
5. Sperm complete their maturation process and become fertile as they move through the _____.
6. Female sex cells, or gametes, develop in the ovaries by a form of _____ called oogenesis.
7. Functionally, the mammary glands produce milk; structurally, they are modified _____.
8. The bubble-like sac that surrounds the baby while it is inside of the mother's uterus: _____.
9. The _____ is the female egg cell.
10. The _____ is the entire outside genital area of a female.
11. Female ovoid structure about 35 cm long, 2 cm wide and 1 cm thick: _____.

References and Recommended Reading

Halperin, L.R. 2004. Study shows female mice produce new eggs. *The Harvard Crimson: On Line.* Accessed January 20, 2007, at www.thecrimson.com/article.

Hyde, J.S. 2005. *Biological Substrate of Human Sexuality.* New York: American Psychological Association.

James, R.E., and K.H. Lopez. 2006. *Human Reproductive Biology,* 3rd ed. New York: Academic Press.

Kimmel, M.S. 2003. *The Gendered Society.* Oxford University Press.

Lips, H.M. 2004. *Sex and Gender: An Introduction.* New York: McGraw-Hill.

Mader, S.S. 2004. *Human Reproductive Biology.* New York: McGraw-Hill.

SEER. 2000. *Reproduction.* Accessed August 19, 2006, at http://training.seer.cancer.gov.

The Universe Within Touring Company. 2007. *Our Body Systems.* Accessible by www.ourbody theuniversewithin.com/about.htm.

CHAPTER 15

Musculoskeletal System

All routine functions, crucial to our daily activity, are dependent on the musculoskeletal system.

Topics in This Chapter

- Human Muscles
- Human Muscle Disorders
- Human Skeleton
- Human Skeletal System
- Human Skeletal Disorders

The **musculoskeletal system** consists of muscles and tendons and ligaments and bones and joints and associated tissues that move the body and maintain form. More specifically, all movements of the human body are dependent on the functioning of muscles. Muscles function to move the body and appendages; they assist with the mechanics of breathing; and they work to move body fluids within the body. When muscles are active, their cells release heat to the body. Muscles also provide form and stability to the body.

The adult human skeletal system, consisting of 206 bones (80 axial and 126 appendicular bones), provides the framework to which muscles attach/support shape. The skeleton also protects sensitive internal organs. Bones provide attachments for muscles (act as levers), which are important in body movements. The marrow from the center of bones produces blood cells. When minerals (e.g., calcium) are not readily available in sufficient amounts, they may be withdrawn from bones.

DEFINITION OF KEY TERMS

- **Aponeurosis**—a whitish, fibrous membrane that connects a muscle to a bone or fascia
- **Appositional growth**—growth accomplished by the addition of new layers on those previously formed
- **Canaliculi**—radiate from the lacunae into surrounding bone matrix
- **Chondrocytes**—produce and maintain the cartilaginous matrix
- **Diaphysis**—the main or mid section (shaft) of a long bone
- **Endomysium**—a layer of connective tissue that ensheaths a muscle fiber and is composed mostly from reticular fibers
- **Epimysium**—a layer of connective tissue which ensheaths the entire muscle
- **Epiphyseal plate**—the growing ends of young bones
- **Fasciculus**—a bundle of fibers
- **Hematopoiesis**—process by which immature precursor cells develop into mature blood cells
- **Lamellae**—portion of compact bone, used in anchoring
- **Osteocytes**—most abundant cell found in bone
- **Perimysium**—is a sheath of connective tissue which groups individual muscle fibers into bundles or fascicles
- **Periosteum**—is an envelope of fibrous connective tissue that is wrapped around the bone in all places except in joints
- **Stapedium muscle**—muscle of the middle ear

Human Muscles (SEER 2000)

The human muscular system is composed of specialized cells called muscle fibers. Their predominant function is contractibility. Muscles, where attached to bones or internal organs and blood vessels, are responsible for movement. Nearly all movement in the body is the result of muscle contraction.

Obvious movements such as walking and running are the results of the integrated action of joints, bones, and skeletal muscles. Moreover, the more subtle movements that result, for example, in various facial expressions, eye movements, and respiration are produced by skeletal muscles.

In addition to movement, muscle contraction also fulfills some other important functions in the body, such as posture, joint stability, and heat production. Sitting and standing (posture) is maintained as a result of contraction. The skeletal muscles are continually making fine adjustments that hold the body in stationary positions. The tendons of any muscles extend over joints and in this way contribute to joint stability. This is particularly evident in the knee and shoulder joints, where muscle tendons are a major factor in stabilizing the joint. In maintaining body temperature, heat is an important by-product of muscle metabolism. In fact, nearly 85% of the heat produced in the body is the result of muscle contraction.

✓   **Important Point**: A whole skeletal muscle is considered an organ of the muscular system. Each organ or muscle consists of skeletal muscle tissue, connective tissue, nerve tissue, and blood or vascular tissue.

SKELETAL MUSCLE: STRUCTURE AND FUNCTION

Skeletal muscles vary considerably in size, shape, and arrangement of fibers. They range from extremely tiny strands such as the *stapedium* muscle of the middle ear to large masses such as the muscles of the thigh. Some skeletal muscles are broad in shape and some narrow. In some muscles the fibers are parallel to the long axis of the muscle, in some they converge to a narrow attachment, and in some they are oblique.

Each skeletal muscle fiber is a single cylindrical muscle cell. An individual skeletal muscle may be made up of hundreds, or even thousands, of muscle fibers bundled together and wrapped in a connective tissue covering. Each muscle is surrounded by a connective tissue sheath called the *epimysium*. Fascia, connective tissue outside the epimysium, surrounds and separates the muscles. Portions of the epimysium project inward to divide the muscle into compartments. Each compartment contains a bundle of muscle fibers. Each bundle of muscle fiber is called a *fasciculus* and is surrounded by a layer of connective tissue called the *perimysium*. Within the fasciculus, each individual muscle cell, called a muscle fiber, is surrounded by connective tissue called the *endomysium*.

Skeletal muscle cells (fibers), like other body cells, are soft and fragile. The connective tissue coverings furnish support and protection for the delicate cells and allow them to withstand the forces of contraction. The coverings also provide pathways for the passage of blood vessels and nerves.

Commonly, the epimysium, perimysium, and endomysium extend beyond the fleshy part of the muscle, the belly or gaster, to form a thick ropelike tendon or a broad, flat sheet-like *aponeurosis*. The tendon and aponeurosis form indirect attachments from muscles to the periosteum of bones or to the connective tissue of other muscles. Typically a muscle spans a joint and is attached to bones by tendons at both ends. One of the bones remains relatively fixed or stable while the other end moves as a result of muscle contraction.

✓   **Important Point**: Each muscle fiber is surrounded by endomysium. The fibers are collected into bundles covered by perimysium. Many bundles, or fasciculi, are wrapped together by the epimysium to form a whole muscle.

Skeletal muscles have an abundant supply of blood vessels and nerves. This is directly related to the primary function of skeletal muscle, contraction. Before a skeletal muscle fiber can contract, it has to receive an impulse from a nerve cell. Generally, an artery and at least one vein accompany each nerve that penetrates the epimysium of a skeletal muscle. Branches of the nerve and blood vessels follow the connective tissue components of the muscle of a nerve cell and with one or more minute blood vessels called capillaries.

✓   **Important Point**: One of the most predominant characteristics of skeletal muscle tissue is its contractility and nearly all movement in the body is the result of muscle contraction.

TYPES OF MUSCLE (SEER 2000)

In the body, there are 3 types of muscle: skeletal (striated), smooth, and cardiac:

- **Skeletal muscle**—attached to bones, is responsible for skeletal movements. The peripheral portion of the central nervous system (CNS) controls the skeletal muscles. Thus, these muscles are under conscious, or voluntary, control. The basic unit is the muscle fiber with many nuclei. These muscle fibers are striated (having transverse streaks) and each acts independently of neighboring muscle fibers.
- **Smooth muscle**—found in the walls of the hollow internal organs such as blood vessels, the gastrointestinal tract, bladder, and uterus; is under control of the autonomic nervous system. Smooth muscle cannot be controlled consciously and thus acts involuntarily. The nonstriated (smooth) muscle cell is spindle-shaped and has 1 central nucleus. Smooth muscle contracts slowly and rhythmically.
- **Cardiac muscle**—found in the walls for the heart, is also under control of the autonomic nervous system. The cardiac muscle cell has 1 central nucleus, like smooth muscle, but also is striated, like skeletal muscle. The cardiac muscle cell is rectangular in shape. The contraction of cardiac muscle is involuntary, strong, and rhythmical.

✓   **Important Point**: Four functions of muscle contraction are movement, posture, joint stability, and heat production.

MUSCLE GROUPS (SEER 2000)

The more than 600 muscles in the human body account for about 40% of a person's weight. Most skeletal muscles have names that describe some feature of the muscle. Often several criteria are combined into one name. The following are some terms relating to muscle features that are used in naming muscles:

- **Size**—vastus (huge); maximus (large); longus (long); minimus (small); brevis (short)
- **Shape**—deltoid (triangular); rhomboid (like a rhombus with equal and parallel sides); latissimus (wide); teres (round); trapezius (like a trapezoid, a 4-sided figure with 2 sides parallel)
- **Direction of fibers**—rectus (straight); transverse (across); oblique (diagonally); orbicularis (circular)
- **Location**—pectorals (chest); gluteus (buttock or rump); brachii (arm); supra- (above); infra- (below); sub- (under or beneath); lateralis (lateral)

- **Number of origins**—biceps (2 heads); triceps (3 heads); quadriceps (4 heads)
- **Origin and insertion**—sternocleidomastoideus (origin on the sternum and clavicle, insertion on the mastoid process); brachioradialis (origin on the brachium or arm, insertion on the radius)
- **Action**—abductor (to abduct a structure); adductor (to adduct a structure); flexor (to flex a structure); extensor (to extend a structure); levator (to lift or elevate a structure); masseter (a chewer)

✓   **Important Point**: Muscles are attached to bones by tendons.

Human Muscle Disorders

Numerous and assorted muscle disorders can occur in the human muscular system. In this text, for illustrative purpose only, 2 of the best-known disorders are discussed.

FIBROMYALGIA

According to the National Institutes of Health (2005), **fibromyalgia** is a disorder that causes muscle pain and fatigue (feeling tired). People with fibromyalgia have "tender points" on the body. Tender points are specific places on the neck, shoulders, back, hips, arms, and legs. These points hurt when pressure is put on them. The causes of fibromyalgia are unknown.

✓   **Important Point**: Fibromyalgia affects as many as 1 in 50 Americans. Most people with fibromyalgia are women. However, men and children also can have the disorder. Most people are diagnosed during middle age.

MUSCULAR DYSTROPHY

The **muscular dystrophies** (MD) are a group of more than 30 genetic diseases characterized by progressive weakness and degeneration of the skeletal muscles that control movement. Some forms of MD are seen in infancy or childhood, while others may not appear until middle age or later. The disorders differ in terms of the distribution and extent of muscle weakness (some form of MD also affect cardiac muscle), age of onset, rate of progression, and pattern of inheritance.

✓   **Important Point**: There is no specific treatment to stop or reverse any form of MD. Treatment may include physical therapy, respiratory therapy, speech therapy, orthopedic appliances used for support and corrective orthopedic surgery.

Human Skeleton

As mentioned, the human skeleton usually consists of 206 named bones. These bones can be grouped in 2 divisions: axial skeleton and appendicular skeleton. The 80 bones of the axial skeleton form the vertical axis of the body. They include the bones of the head, vertebral column, ribs, and breastbone or sternum. The appendicular skeleton consists of 126 bones and includes the free appendages and their attachments to the axial skeleton. The free appendages are the upper and lower extremities, or limbs, and their attachments (called girdles). The named bones of the body are listed below by category.

AXIAL SKELETON (80 BONES)

Skull (28)

Cranial Bone
- Parietal (2)
- Temporal (2)
- Frontal (1)
- Occipital (1)
- Ethmoid (1)
- Sphenoid (1)

Facial Bones
- Maxilla (2)
- Zygomatic (2)
- Mandible (1)
- Nasal (2)
- Platine (2)
- Inferior nasal concha (2)
- Lacrimal (2)
- Vomer (1)

Auditory Ossicles
- Malleus (2)
- Incus (2)
- Stapes (2)

Hyoid (1)

Vertebral Column (26)
- Cervical vertebrae (7)
- Thoracic vertebrate (12)

- Lumbar vertebrae (5)
- Sacrum (1)
- Coccyx (1)

Thoracic Cage (25)
- Sternum (1)
- Ribs (24)

APPENDICULAR SKELETON (126 BONES)

Pectoral Girdles (4)

- Clavicle (2)
- Scapula (2)

Upper Extremity (60)

- Humerus (2)
- Radius (2)
- Ulna (2)
- Carpals (16)
- Metacarpals (10)
- Phalanges (28)

Pelvic Girdle

- Coxal, innominate, or hip bones (2)

Lower Extremity

- Femur (2)
- Tibia (2)
- Fibula (2)
- Patella (2)
- Tarsals (14)
- Metatarsals (10)
- Phalanges (28)

Human Skeletal System (SEER 2000)

Humans rely on a sturdy internal frame that is centered on a prominent spine. The human skeletal system consists of bones, cartilage, ligaments, and tendons, and accounts for about 20% of the body weight. The living bones in the body use oxygen and give off waste products in metabolism. They contain active tissues that consume nutrients, require a blood supply, and change shape or remodel in response to variations in mechanical stress.

Bones provide a rigid framework (the skeleton) that supports and protects the soft organs of the body. The skeleton supports the body against the pull of gravity. The large bones of the lower limbs support the trunk when standing. The fused bones of the cranium surround the brain to make it less vulnerable to injury. Vertebrae surround and protect the spinal cord and bones of the rib cage help protect the heart and lungs of the thorax.

✓   **Important Point**: Bones work together with muscles as simple mechanical lever systems to produce body movement.

Bones contain more calcium than any other organ. The intercellular matrix of bone contains large amounts of calcium salts, the most important being calcium phosphate. When the blood calcium levels decrease below normal, calcium is released from the bones so that there will be an adequate supply for metabolic needs. When blood calcium levels are increased, the excess calcium is stored in the bone matrix. The dynamic process of releasing and storing calcium goes on almost continuously.

✓   **Important Point**: **Hematopoiesis**, the formation of blood cells, mostly takes place in the red marrow of the bones.

In infants, red marrow is found in the bone cavities. With age, it is largely replaced by yellow marrow for fat storage. In adults, red marrow is limited to the spongy bone in the skull, ribs, sternum, clavicles, vertebrae, and pelvis. Red marrow functions in the formation of red blood cells, white blood cells, and blood platelets.

✓   **Important Point**: The human skeleton is well adapted for the functions it must perform. Functions of bones include support, protection, movement, mineral storage, and formation of blood cells.

TYPES OF BONE TISSUE (SEER 2000)

There are 2 types of bone tissue: compact and spongy. The names imply that the 2 types differ in density, or how tightly the tissue is packed together. There are 3 types of cells that contribute to bone homoeostasis. Osteoblasts are bone-forming cells, osteoclasts break down bone, and osteocytes are mature bone cells. Equilibrium between osteoblasts and osteoclasts maintains bone tissue.

- **Compact bone**—consists of closely packed osteons or haversian systems. The osteon consists of a central canal called the osteonic (haversian) canal, which is surrounded by concentric rings (lamellae) of matrix. Between the rings of matrix, the cone cells (osteocytes) are located in spaces called lacunae. Small channels (canaliculi) radiate from the lacunae to the osteonic (haversian) canal to provide passage-ways through the hard matrix. In compact bone, the haversian systems are packed tightly together to form what appears to be a solid mass. The osteonic canals contain blood vessels that are parallel to the long axis of the bone. These blood vessels interconnect, by way of perforating canals, with vessels on the surface of the bone.

- **Spongy (cancellous) bone**—is lighter and less dense than compact bone. Spongy bone consists of plates (trabeculae) and bars of bone adjacent to small, irregular cavities that contain red bone marrow. The canaliculi connect to the adjacent cavities, instead of a central haversian canal, to receive their blood supply. It may appear that the trabeculae are arranged in a haphazard manner, but they are organized to provide maximum strength, similar to braces that are used to support a building. The trabeculae of spongy bone follow the lines of stress and can realign if the direction of stress changes.

BONE DEVELOPMENT (SEER 2000)

The terms *osteogenesis* and *ossification* are often used synonymously to indicate the process of bone formation. Parts of the skeleton form during the first few weeks after conception. By the end of the 8th week after conception, the skeletal pattern is formed in cartilage and connective tissue membranes and ossification begins.

Bone development continues throughout adulthood. Even after adult stature is attained, bone development continues for repair of fractures and for remodeling to meet changing lifestyles. Osteoblasts, osteocytes and osteoclasts are the 3 cell types involved in the development, growth and remodeling of bones. Osteoblasts are bone-forming cells, osteocytes are mature bone cells and osteoclasts break down and reabsorb bone.

There are 2 types of ossification, intramembranous and endochondral:

- **Intramembranous ossification**—involves the replacement of sheet-like connective tissue membranes with bony tissue. Bones formed in this manner are called intramembranous bones. They include certain flat bones of the skull and some of the irregular bones.
- **Endochondral ossification**—involves the replacement of hyaline cartilage with bony tissue. Most of the bones of the skeleton are formed in this manner.

BONE GROWTH (SEER 2000)

Bones grow in length at the epiphyseal plate by a process that is similar to endochondral ossification. The cartilage in the region of the epiphyseal plate next to the epiphysis continues to grow by mitosis. The chondrocytes, the region next to the diaphysis, age and degenerate. Osteoblasts move in and ossify the matrix to form bone. This process continues throughout childhood and the adolescent years until the cartilage growth slows and finally stops. When cartilage growth ceases, usually in the early 20s, the epiphyseal plate completely ossifies so that only a thin epiphyseal line remains and the bones can no longer grow in length. Bone growth is under the influence of growth hormone from the anterior pituitary gland and sex hormones from the ovaries and testes.

✓   **Important Point**: Even though bones stop growing in length in early adulthood, they can continue to increase in thickness or diameter throughout life in response to stress from increased muscle activity or to weight. The increase in diameter is called appositional growth.

CLASSIFICATION OF BONES

The bones of the human body come in a variety of sizes and shapes. The 4 principal types of bones are long, short, flat, and irregular:

- **Long bones**—are longer than they are wide. They consist of a long shaft with 2 bulky ends or extremities. They are primarily compact bone but may have a large amount of spongy bone at the ends or extremities. Long bones include bones of the thigh, leg, arm, and forearm.
- **Short bones**—are roughly cube shaped with vertical and horizontal dimensions approximately equal. They consist primarily of spongy bone, which is covered by a thin layer of compact bone. Short bones include the bones of the wrist and ankle.
- **Flat bones**—are thin, flattened, and usually curved. Most of the bones of the cranium are flat bones.
- **Irregular bones**—includes those bones not listed in the 3 categories above. They are primarily spongy bone that is covered with a thin layer of compact bone. The vertebrae and some of the bones in the skull are irregular bones.

✓   **Important Point**: All bones have surface markings and characteristics that make a specific bone unique. There are holes, depressions, smooth facets, lines, projections, and other markings. These usually represent passageways for vessels and nerves, points of articulation with other bones, or points of attachment for tendons and ligaments.

ARTICULATIONS (SEER 2000)

An articulation, or joint, is where 2 bones come together. In terms of the amount of movement they allow, there are 3 types of joints: immovable, slightly movable and freely movable:

- **Immovable joints** (*synarthroses*; singular, *synarthrosis*)—in these joints, the bones come in very close contact and are separated only by a thin layer of fibrous connective tissue. The sutures in the skull are examples of immovable joints.

- **Slightly movable joints** (*amphiarthroses*; singular, *amphiarthrosis*)—in this type of joint, the bones are connected by hyaline cartilage or fibrocartilage. The ribs connecte to the sternum by costal cartilages are slightly movable joints connected by hyaline cartilage. The symphysis pubis is a slightly movable joint in which there is a fibrocartilage pad between the 2 bones. The joints between the vertebrae and the intervertebral disks are also of this type.
- **Freely movable joints** (*diarthroses*; singular, *diarthrosis*)—most joints in the adult human body are of this type. In this type of joint, the ends of the opposing bones are covered with hyaline cartilage, the articular cartilage, and they are separated by a space called the joint cavity. The components of the joints are enclosed in a dense fibrous joint capsule. The outer layer of the capsule consists of the ligaments that hold the bones together. The inner layer is the synovial membrane that secretes synovial fluid into the joint cavity for lubrication. Because all of these joints have a synovial membrane, they are sometimes called synovial joints.

Skeletal System Disorders

As with muscular disorders, skeletal disorders are numerous and assorted. For illustrative purposes, 3 of the most common and well-known disorders, bone fracture, osteoporosis, and vitamin D deficiency, are discussed in the following.

BONE FRACTURES

A **bone fracture** (broken bone) occurs when a force exerted against bone is stronger than the bone can structurally withstand. Bones are a form of connective issue, reinforced with calcium and bone cells. Bones have a softer center, called marrow, where blood cells are made. As mentioned, the main functions of the skeleton include support, movement and protection of vulnerable internal organs. There are different types of bone fractures that vary in severity. Factors that influence severity include the degree and direction of the forces, the particular bone involved, and the person's age and general health. Common sites for bone fractures include the wrist, ankle and hip. Hip fractures occur most often in elderly people. Broken bones take around 4 to 8 weeks to heal, depending on the age, health of the individual, and the type of break.

TYPES OF BONE FRACTURE

The different types of bone fracture include:

- **Greenstick fracture**—the bone sustains a small, slender crack. This type of fracture is more common in children, due to the comparative flexibility of their bones.
- **Comminuted fracture**—the bone is shattered into small pieces. This type of complicated fracture tends to heal at a slower rate.

- **Simple fracture** (or *closed fracture*)—the broken bone has pierced the skin.
- **Compound fracture** (or *open fracture*)—the broken bone juts through the skin, or a wound leads to the fracture site. The risk of infection is higher with this type of fracture.
- **Pathological fracture**—bones weakened by various diseases (such as osteoporosis or cancer) tend to break with very little force.
- **Avulsion fracture**—muscles are anchored to bone with tendons, a type of connective tissue. Powerful muscle contractions can wrench the tendon free, and pull out pieces of bone. This type of fracture is more common in the knee and shoulder joints.
- **Compression fracture**—occurs when 2 bones are forced against each other. The bones of the spine, called vertebrae, are prone to this type of fracture. Elderly people, particularly those with osteoporosis, are at increased risk.

VITAMIN D DEFICIENCY

Vitamin D is a fat-soluble vitamin that is found in food and can also be made in your body after exposure to ultraviolet (UV) rays from the sun. Sunshine is a significant source of vitamin D because UV rays from sunlight trigger vitamin D synthesis in the skin.

The major biologic function of vitamin D is to maintain normal blood levels of calcium and phosphorus. By promoting calcium absorption, vitamin D helps to form and maintain strong bones. Vitamin D also works in concert with a number of other vitamins, minerals, and hormones to promote bone mineralization. Without vitamin D, bones can become thin, brittle, or misshapen. Vitamin D sufficiency prevents rickets in children and osteomalacia in adults, 2 forms of skeletal diseases that weaken bones.

✓   **Important Point:** Nutrient deficiencies are usually the result of dietary inadequacy, impaired absorption and utilization, increased requirement, or increased excretion (loss).

Chapter Review Questions

1. Most movement in the human body is the result of _____.
2. In addition to movement, muscle contraction also fulfills some other important functions in the body, such as _____, _____, and _____.
3. Each organ or muscle consists of skeletal muscle tissue, connective tissue, nerve tissue, and blood or vascular tissue.

   - True
   - False

4. Tendon is formed by _____, _____, and _____.

5. Cardiac muscle, found in the walls of the heart, is under control of the _____ nervous system.

6. The skeleton protects the _____.

7. When blood calcium levels decrease below normal, _____ is released from the bones so that there will be an adequate supply for metabolic needs.

8. The osteon consists of a central canal called the _____, which is surrounded by concentric rings of matrix.

9. Even though bones stop growing in length in early adulthood, they can continue to increase in thickness or diameter throughout life in response to stress from increased muscle activity or weight.

   • True
   • False

10. Short bones consist primarily of _____, which is covered by a thin layer of compact bone.

References and Recommended Reading

BetterHealthChannel. 2006. *Bone Fractures*. Victoria, Australia. Accessed September 9, 2006, at www.betterhealth.vic.gov.au/bhcv2/bhcarticles.nsf/pages/Bone_fractures?open.

Chesney, R. 2003. Rickets: An Old Form for a New Century. *Pediatrics International* 45:509–11.

Gaby, A.R.1994. *Preventing and Reversing Osteoporosis*. New York: Random House.

NIH. 2005. *What Is Fibromyalgia?* National Institutes of Health. Accessed September 2006, at www.niams.nih.gov/hi/topics/fibromyalgia/fffibro.htm.

NIH. 2006. *Dietary Supplement Use & Safety*. Accessed September 9, 2006, at http://dietary-supplements.info.nih.gov.

SEER. 2000. *Anatomy & Physiology: Skeletal and Muscular System*. Accessed September 6, 2006, at http://training.seer.cancer.gov/module.

Van den Berg, H. 1997. Bioavailability of Vitamin D. *European Journal of Clinical Nutrition* 51 Suppl 1:S76-9.

U.S. Department of Agriculture, Agricultural Research Service. 2003. *USDA Nutrient Database for Standard Reference, Release 16*. Nutrient Data Laboratory Home Page. Accessed August 18, 2006, at www.nal.usda.gov/fnic.foodcomp.

CHAPTER 16

Respiratory System

I know that our bodies were made to thrive only in pure air, and the scenes in which pure air is found.

—John Muir (1838–1914)

Topics in This Chapter

- Mechanics of Breathing
- Respiratory Volumes and Capacities
- Respiratory Conducting Passages
- Respiratory System Disorders

When the **respiratory system** is mentioned, we generally think of breathing, but breathing is only one of the activities of the respiratory system. Respiration involves the complex exchange of gases. Air is inhaled through the mouth and nose and carried to tiny air sacs known as the alveoli. In the alveoli, oxygen diffuses into blood and carbon dioxide diffuses out. This carbon dioxide is then carried back up the airways and exhaled. This process of gas exchange provides all cells the oxygen so essential for life and eliminates carbon dioxide, the main end product of metabolism. It also helps to regulate pH of the blood.

Respiration is the sequence of events that results in the exchange of oxygen and carbon dioxide between the atmosphere and the body cells. Every 3 to 5 seconds, nerve impulses stimulate the breathing process, or ventilation, which moves air through a series of passages into and out of the lungs. After this, *external respiration* occurs; that is, there is an air exchange of gases between the lungs and the blood. *Internal respiration* occurs when the blood transports the gases to and from the tissue cells. Finally, the cells utilize the oxygen for their specific activities. This is *cellular respiration*. Together these activities constitute respiration (SEER 2000).

DEFINITION OF KEY TERMS

- **Bronchi**—the large air tubes leading from the trachea to the lungs that convey air to and from the lungs
- **Diaphragm**—a shelf of muscle extending between the thorax and abdomen
- **Ethmoid**—bone in the skull that separates the nasal cavity from the brain
- **Expired reserve volume (ERV)**—maximal volume that can be exhaled after the exhalation of a tidal volume/normal breath
- **Functional residual capacity (FRC)**—volume of gas remaining in lung after normal expiration
- **Inhaled reserve volume (IRV)**—maximum volume that can be inhaled over the inhalation of a tidal volume/normal breath; used during exercise/exertion
- **Inspiratory capacity (IC)**—volume of maximal inhalation
- **Maxillae**—the largest bones of the face except for the mandible, which, by their union, form the whole upper jaw
- **Mediastinum**—a nondelineated group of structures in the chest, surrounded by loose connective tissue
- **Pleura**—a thin serous membrane that envelops each lung and folds back to make a lining from the chest cavity
- **Residual volume (RV)**—volume that remains in the lungs after a maximal expiration
- **Sphenoid**—a butterfly-shaped bone at the base of the skull
- **Total lung capacity (TLC)**—the volume of the lung after maximal inhalation
- **Total volume (TV)**—volume inhaled or exhaled with each normal breath
- **Ventilation**—the rate at which gas enters or leaves the lung
- **Vital capacity (VC)**—volume of maximal inhalation and exhalation

Mechanics of Breathing

Breathing is the movement of air through the conducting passages between the atmosphere and the lungs. The air moves through the passages because of pressure gradients that are produced by contraction of the diaphragm and thoracic muscles.

PULMONARY VENTILATION (SEER 2000)

Pulmonary ventilation is commonly referred to as breathing. It is the process of air flowing into the lungs during inhalation (inspiration) and out of the lungs during exhalation (expiration). Air flows because of pressure differences between the atmosphere and the gases inside the lungs.

Like other gases, air flows from a region with higher pressure to a region with lower pressure. Muscular breathing movements and recoil of elastic tissues create the changes in pressure that result in ventilation. Pulmonary ventilation involves 3 different pressures, all of which are responsible for pulmonary ventilation:

- **Atmospheric pressure**—the pressure of the air outside the body
- **Intraalveolar pressure**—the pressure inside the alveoli of the lungs
- **Intrapleural pressure**—the pressure within the pleural cavity

Inhalation is the process of taking air into the lungs. It is the active phase of ventilation because it is the result of muscle contraction. During inhalation, the diaphragm contracts and the thoracic cavity increases in volume. This decreases the intraalveolar pressure so that air flows into the lungs. Inhalation draws air into the lungs.

Exhalation is the process of letting air out of the lungs during the breathing cycle. During exhalation, the relaxation of the diaphragm and elastic recoil of tissue decrease the thoracic volume and increases the intraalveolar pressure. Exhalation pushes air out of the lungs.

Respiratory Volumes and Capacities

The average adult, under normal conditions, takes 12 to 15 breaths a minute. A breath is 1 complete respiratory cycle that consists of 1 inhalation and 1 exhalation.

In a process called spirometry (pulmonary function testing, PFT), a spirometer is used to measure the volume of air that moves into and out of the lungs. Respiratory (pulmonary) volumes are an important aspect of pulmonary function testing (PFT) because they can provide information about the physical condition of the lungs. Factors such as age, sex, body build, and physical conditioning have an influence on lung volumes and capacities. Lungs usually reach their maximum capacity in early adulthood and decline with age after that (SEER 2000).

✓   **Important Point**: Pulmonary capacity is the sum of 2 or more volumes.

Respiratory Conducting Passages (SEER 2000)

The respiratory conducting passages are divided into the upper respiratory tract and the lower respiratory tract. The upper respiratory tract includes the nose, pharynx, and larynx. The lower respiratory tract consists of the trachea, bronchial tree, and lungs. These tracts open to the outside and are lined with mucous membranes. In some regions, the membrane has hairs that help filter the air. Other regions may have cilia to propel mucus.

NOSE AND NASAL CAVITIES

The nose has been called the "air conditioner of the body" (BCM 2006). The nose is responsible for warming and saturating inspired air, removing bacteria and particulate matter, as well as conserving heat and moisture from exhaled air. Nasal breathing is important for optimal pulmonary function.

The framework of the nose consists of bone and cartilage. Two small nasal bones and extensions of the maxillae form the bridge of the nose, which is the bony portion. The remainder of the framework is cartilage and is the flexible portion. Connective issue and skin cover the framework.

✓   **Important Point**: Air enters the nasal cavity from the outside through 2 openings, the nostrils, or external nares. The openings from the nasal cavity into the pharynx are the internal nares. Nose hairs at the entrance to the nose trap large inhaled particles.

Paranasal Sinuses

Paranasal sinuses are air-filled cavities in the frontal, maxilla, ethmoid, and sphenoid bones. These sinuses, which have the same names as the bones in which they are located, surround the nasal cavity and open into it. They function to reduce the weight of the skull, to produce mucus, and to influence voice quality by acting as resonating chambers.

PHARYNX

The **pharynx** (from Greek, meaning "throat") is a passageway that extends from the base of the skull to the level of the 6th cervical vertebra. It serves both the respiratory and digestive systems by receiving air from the nasal cavity and air, food, and water from the oral cavity. Inferiorly, it opens into the larynx and esophagus. The pharynx is divided into 3 regions according to location: the nasopharynx, the oropharynx, and the laryngopharynx.

The **nasopharynx** is the portion of the pharynx that is posterior to the nasal cavity and extends inferiorly to the uvula. The **oropharynx** is the portion of the pharynx that is posterior to the oral cavity. The most inferior portion of the pharynx is the **laryngopharynx** that extends from the hyoid bone down to the lower margin of the larynx.

✓   **Important Point**: The upper part of the pharynx lets only air pass through. Lower parts permit air, foods, and fluids to pass.

LARYNX

The **larynx**, or voice box, is the passageway for air between the pharynx above and the trachea below. It extends from the 4th to the 6th vertebral levels. The larynx is often divided into 3 sections: sublarynx, larynx, and supralarynx. It is formed by 9 cartilages that are connected to each other by muscles and ligaments.

The larynx plays an essential role in human speech. During sound production, the vocal cords close together and vibrate as air expelled from the lungs passes between them. The false vocal cords have no role in sound production, but help close off the larynx when food is swallowed.

 Important Point: The thyroid cartilage is the Adam's apple. The epiglottis acts like a trap door to keep food and other particles from entering the larynx.

TRACHEA

The **trachea**, or windpipe, is the main airway to the lungs. It divides into the right and left bronchi at the level of the 5th thoracic vertebra, channeling air to the right or left lung.

- **Hyaline cartilage** in the tracheal wall provides support and keeps the trachea from collapsing. The posterior soft tissue allows for expansion of the esophagus, which is immediately posterior to the trachea.
- **Mucous membrane** that lines the trachea is ciliated pseudostratified columnar epithelium, similar to that in the nasal cavity and nasopharynx.
- **Goblet cells** produce mucus that traps airborne particles and microorganisms, and the cilia propel the mucus upward, where it is either swallowed or expelled.

BRONCHI AND BRONCHIAL TREE

The trachea, in the mediastinum, divides into the right and left primary **bronchi**. The bronchi branch into smaller and smaller passageways until they terminate in tiny air sacs called *alveoli*.

The cartilage and mucous membrane of the primary bronchi are similar to that in the trachea. As the branching continues through the **bronchial tree**, the amount of hyaline cartilage in the walls decreases until it is absent in the smallest bronchioles. As the cartilage decreases, the amount of smooth muscle increases. The mucous membrane also undergoes a transition from ciliated pseudostratified columnar epithelium to simple cuboidal epithelium to simple squamous epithelium.

 Important Point: The alveolar ducts and alveoli consist primarily of simple squamous epithelium, which permits rapid diffusion of oxygen and carbon dioxide. Exchange of gases between the air in the lungs and the blood in the capillaries occurs across the walls of the alveolar ducts and alveoli.

LUNGS

Most of the thoracic cavity is occupied by the 2 lungs, which contain all the components of the bronchial tree beyond the primary bronchi. The lungs are soft and spongy because they are mostly air spaces surrounded by the alveolar cells and elastic connective tissue. They are separated from each other by the mediastinum, which contains the heart. The only point of attachment for each lung is at the hilum, or root, on the medial side. This is where the bronchi, blood vessels, lymphatics, and nerves enter the lungs.

The right lung is shorter, broader, and has a greater volume than the left lung. It is divided into 3 lobes and each lobe is supplied by one of the secondary bronchi. The left lung is longer and narrower than the right lung. It has an indentation, called the cardiac notch, on its medial surface for the apex of the heart. The left lung has 2 lobes.

✓   **Important Point**: The right lung is shorter, broader, and it is divided into 3 lobes. The left lung is longer, narrower, and it is divided into 2 lobes.

Each lung is enclosed by a double-layered serous membrane, called the *pleura*. The visceral pleura is firmly attached to the surface of the lung. At the hilum, the visceral pleura is continuous with the parietal pleura that lines the wall of the thorax. The small space between the visceral and parietal pleurae is the pleural cavity. It contains a thin film of serous fluid that is produced by the pleura. The fluid acts as a lubricant to reduce friction as the 2 layers slide against each other, and it helps to hold the 2 layers together as the lungs inflate and deflate.

Respiratory System Disorders

Some of the disorders of the respiratory system are discussed below:

- **Asthma**—is an allergic response resulting in the constriction of the bronchial tubes. Simply, asthma is a chronic disease, affecting more than 10 million Americans, that makes it hard to breathe. Deaths due to asthma have increased dramatically in recent years (ALA 2005). Asthma attacks can be triggered by pollens, dust mites, animal dander, smoke, cold air, and exercise.
- **Emphysema**—occurs when the airway walls lose their structural support (elasticity; ability to stretch and recoil). Air sacs (alveoli) become weakened and break. When elasticity is lost, air is trapped in the air sacs and impairs the exchange of oxygen and carbon dioxide. Symptoms of emphysema include shortness of breath, cough, and a limited exercise tolerance. The result is difficulty in breathing, overwork of the heart, and very often death. Emphysema doesn't develop suddenly, but gradually; it is insidious, like most cancers.
- **Chronic obstructive pulmonary disease (COPD)**—is a general term for a group of diseases that cause progressive damage to the lungs. These diseases include chronic bronchitis, asthma, and emphysema.
- **Lung cancer**—occurs when a tumor forms in the tissue of the lung. Lung cancer is the leading cause of cancer death in men and women in the United States (NCI 2006). Cancers frequently spread to the lungs from other parts of the body by be-

ing carried through the bloodstream. Most lung cancers are carcinomas, a cancer that begins in the lining or covering tissues of an organ.

- **Pleurisy**—is the inflammation of the pleural linings of the lungs, caused by an accumulation of fluid between the pleural layers.
- **Bronchitis**—is inflammation of the membranes that line the bronchial tubes.
- **Hiccup**—is irritation of the nerves that control the diaphragm, resulting in an irregular intake of air which causes a peculiar noise as the glottis closes down.

Chapter Review Questions

1. The exchange of gases between the blood and tissue cells is _____ respiration.
2. Air flows in the process of pulmonary ventilation because of pressure differences between the _____ and the _____ inside the lungs.
3. The amount of air that is exchanged during 1 cycle varies with _____, _____, _____, and physical condition.
4. External _____ are 2 openings through which air enters the nasal cavity from the outside.
5. The upper part of the _____ lets only air pass through. Lower parts permit air, foods and fluids to pass.
6. The right lung and the left lung are the same in size.

   - True
   - False

7. _____ is a group of diseases that includes chronic bronchitis, asthma, and emphysema.
8. The trachea divides into the right and left _____.
9. The vocal cords have no role in sound production, but help close off the larynx when food is swallowed.

   - True
   - False

10. _____ is commonly referred to as breathing.

References and Recommended Reading

American Lung Association (ALA). 2006. *Trends in Asthma Morbidity and Mortality*. Accessed September 9, 2006, at www.lungusa.org.
American Lung Association (ALA). 2006. *What Is Emphysema?* Accessible at www.lungusa.org.

BCM. 2006. *Nose and Paranasal Sinuses.* Bobby R. Alford Department of Otolaryngology. Houston, TX: Baylor College of Medicine.

Canadian Lung Association. 2006. *Asthma.* Accessible at www.lung.ca/diseases.

Mayo Clinic. 2006. *Emphysema.* Accessible at www.mayoclinic.com/health/emphysema.

Muir, J. 1992. *John Muir. The Eight Wilderness Discovery Books.* Luxembourg: Diadem Books.

NCI. 2006. *Lung Cancer Prevention.* Accessed September 10, 2006, at www.cancer.gov.

NIH. 2006. *Asthma.* National Institutes of Health. Accessible at www.nhlbi.nih.gov/health/dci.

SEER. 2000. *Respiratory System.* Accessed September 9, 2006, at http://www.training.eer.concer.gov.

CHAPTER 17

Urinary System

All human cells bathe in an internal sea of fluid with which they exchange nutrients and other molecules essential for life.

—Rita Mary King

Topics in This Chapter

- Functions of the Urinary System
- Components of the Urinary System
- Urinary System Disorders

The human **urinary system**, along with the lungs and the skin, works to expel the wastes produced in metabolic activities. The organs, tubes, muscles, and nerves that work together to create, store, and carry urine are the urinary system. The urinary system includes 2 kidneys, 2 ureters, the bladder, 2 sphincter muscles, and the urethra.

DEFINITION OF KEY TERMS

- **Afferent arteriole**—carries blood into the glomerulus of the nephron.
- **Bowman's (glomerular) capsule**—a sac where fluids from blood in the glomerulus are collected and further processed to form urine.
- **Calyx**—propels urine through the pelvis and ureters to the bladder.
- **Efferent arteriole**—carries blood away from the glomerulus.
- **Glomerulus**—a small cluster or mass of blood vessels or nerves that filters waste products form the blood and this initiates urine formation.
- **Juxtaglomerular apparatus**—a renal structure that is the site of rennin secretion.
- **Loop of Henle**—portion of the nephron that leads from the proximal convoluted tubule to the distal convoluted tubule. Its primary function is to reabsorb water and ions from the urine.

- **Macula densa**—an area of closely packaged cells lining the distal convoluted tubule.
- **Proximal convoluted tubule**—the most proximal segment of the renal tubular portion of the nephron. It is responsible for the reabsorption of glucose, amino acids, various ions and water.
- **Renal capsule**—a tough fibrous layer surrounding the kidney. It provides some protection from trauma and damage.
- **Renal pelvis**—the funnel-like dilated proximal part of ureter. It acts as a funnel for urine flowing to the ureter.
- **Renal pyramids**—cone-shaped tissues of the kidney.
- **Ureter**—long, narrow duct that conveys urine from the kidney to the urinary bladder.

Functions of the Urinary System (SEER 2006)

The principal function of the urinary system is to maintain the volume and composition of body fluids within normal limits. One aspect of this function is to rid the body of waste products that accumulate as a result of cellular metabolism. Because of this, it is sometimes referred to as the *excretory system*.

In addition to the urinary system's major role in excretion, it is important to point out that other organs contribute to the excretory function. For example, the lungs in the respiratory system excrete some waste products, such as carbon dioxide and water. The skin is another excretory organ that rids the body of wastes through the sweat glands. The liver and intestines excrete bile pigments that result from the destruction of hemoglobin. The major task of excretion still belongs to the urinary system. If it fails, the other organs cannot take over and compensate adequately.

By regulating the amount of water that is excreted in the urine, the urinary system maintains an appropriate fluid volume. Other aspects of its function include regulating the concentrations of various electrolytes in the body fluids and maintaining normal pH of the blood.

✓   **Important Point**: In addition to maintaining fluid homeostasis in the body, the urinary system controls red blood cell production by secreting the hormone erythropoietin. The urinary system also plays a role in maintaining normal blood pressure by secreting the enzyme rennin.

Components of the Urinary System (SEER 2006)

The urinary system consists of the kidney, ureters, urinary bladder, and urethra. The kidneys form the urine and account for the other functions attributed to the urinary system. The ureters carry the urine away from kidneys to the urinary bladder, which is a temporary reservoir for the urine. The urethra is a tubular structure that carries the urine from the urinary bladder to the outside.

Important Point: Normal urine is sterile. It contains fluids, salts, and waste products, but is free of bacteria, viruses, and fungi.

KIDNEYS (SEER 2006)

The **kidneys** are the primary organs of the urinary system. The kidneys are the organs that filter the blood, remove the wastes, and excrete the wastes in the urine. They are the organs that perform the functions of the urinary system. The other components are accessory structures to eliminate the urine from the body.

The paired kidneys are located between the 12th thoracic and 3rd lumbar vertebrae, 1 on each side of the vertebral column. The right kidney usually is slightly lower than the left because the liver displaces it downward. The kidneys, protected by the lower ribs, lie in shallow depressions against the posterior abdominal wall and behind the parietal peritoneum. This means they are retroperitoneal. Each kidney is held in place by connective tissue, called renal fascia, and is surrounded by a thick layer of adipose tissue, called perirenal fat, which helps to protect it. A tough, fibrous, connective tissue renal capsule closely envelops each kidney and provides support for the soft tissue that is inside.

 Important Point: In the adult, each kidney is approximately 3 cm thick, 6 cm wide, and 12 cm long. It is roughly bean-shaped with an indentation, called the hilum, on the medial side. The hilum leads to a large cavity, called the renal sinus, within the kidney. The ureter and renal vein leave the kidney, and the renal artery enters the kidney at the hilum.

Each kidney contains over a million functional units, called **nephrons**, in the parenchyma (cortex and medulla). A nephron has 2 parts: a renal corpuscle and a renal tubule. The renal corpuscle consists of a cluster of capillaries, called the glomerulus, surrounded by a double-layered epithelial cup, called the glomerular capsule. An afferent arteriole leads into the renal corpuscle and an efferent arteriole leaves the renal corpuscle. Urine passes from the nephrons into collecting ducts then into the minor calyces.

 Important Point: The **juxtaglomerular apparatus**, which monitors blood pressure and secretes rennin, is formed from modified cells in the afferent arteriole and the ascending limb of the nephron loop.

URETERS

Each **ureter** is a small tube, about 25 cm long, that carries urine from the renal pelvis to the urinary bladder. It descends from the renal pelvis, along the posterior abdominal wall, behind the parietal peritoneum, and enters the urinary bladder on the posterior inferior surface (SEER 2006).

URINARY BLADDER

The **urinary bladder** is a temporary storage reservoir for urine. It is located in the pelvic cavity, posterior to the symphysis pubis, and below the parietal peritoneum. The size and shape of the urinary bladder vary with the amount of urine it contains and with pressure it receives from surrounding organs.

✓   **Important Point**: Adults pass about a quart and a half of urine each day, depending on the fluids and foods consumed.

The inner lining of the urinary bladder is a mucous membrane of transitional epithelium that is continuous with that in the ureters. When the bladder is empty, the mucosa has numerous folds called rugae. The rugae and transitional epithelium allow the bladder to expand as it fills (SEER 2006).

URETHRA

The final passageway for the flow of urine is the **urethra**, a thin-walled tube that conveys urine from the floor of the urinary bladder to the outside. The opening to the outside is the external urethral orifice. The mucosal lining of the urethra is transitional epithelium. The wall also contains smooth muscle fibers and is supported by connective tissue.

The internal urethral sphincter surrounds the beginning of the urethra, where it leaves the urinary bladder. This sphincter is smooth (involuntary) muscle. Another sphincter, the external urethral sphincter, is skeletal (voluntary) muscle and encircles the urethra where it goes through the pelvic floor. These 2 sphincters control the flow of urine through the urethra.

✓   **Important Point**: The flow of urine through the urethra is controlled by an involuntary internal urethral sphincter and voluntary external urethral sphincter.

In females, the urethra is short, only 3 to 4 cm (1.5 inches) long. The external urethral orifice opens to the outside just anterior to the opening for the vagina. In males, the urethra is much longer, about 20 cm (7 to 8 inches) in length, and transports both urine and semen. The first part, next to the urinary bladder, passes through the prostate gland and is called the prostatic urethra. The second part, a short region that penetrates the pelvic floor and enters the penis, is called the membranous urethra. The third part, the spongy urethra, is the longest region. This portion of the urethra extends the entire length of the penis, and the external urethral orifice opens to the outside at the tip of the penis (SEER 2006).

Urinary System Disorders (NIH 2004)

Problems in the urinary system can be caused by aging, illness, or injury. As we get older, changes in the kidneys' structure cause them to lose some of their ability to remove wastes from the blood. Also, the muscles in our ureters, bladder, and urethra tend to lose some of their strength. We may have more urinary infections because the bladder muscles do not tighten enough to empty our bladder completely. A decrease in strength of muscles of the sphincters and the pelvis can also cause incontinence, the unwanted leakage of urine. Illness or injury can also prevent the kidneys from filtering the blood completely or block the passage of urine.

The National Institutes of Health (2004) points out that disorder of the urinary system range in severity from easy-to-treat to life-threatening:

- **Kidney stones**—is the term commonly used to refer to stones, or calculi, in the urinary system. Stones form in the kidneys and may be found anywhere in the urinary system. They vary in size. Some stones cause great pain while others cause very little. The aim of treatment is to remove the stones, prevent infection, and prevent recurrence. Both nonsurgical and surgical treatments are used. Kidney stones affect men more often than women.
- **Benign prostatic hyperplasia (BPH)**—is a condition in men that affects the prostate gland, which is part of the male reproductive system. The prostate is located at the bottom of the bladder and surrounds the urethra. BPH is an enlargement of the prostate gland that can interfere with urinary function in older men. It causes blockage by squeezing the urethra, which can make it difficult to urinate. Men with BPH frequently have other bladder symptoms including an increase in frequency of bladder emptying both during the day and at night. Most men over age 60 have some BPH, but not all have problems with blockage. There are many different treatment options for BPH.
- **Interstitial cystitis (IC)**—is a chronic bladder disorder also known as painful bladder syndrome. In this disorder, the bladder wall can become inflamed and irritated. The inflammation can lead to scarring and stiffening of the bladder, decreased bladder capacity, pinpoint bleeding, and, in rare cases, ulcers in the bladder lining. The cause of IC is unknown at this time.
- **Prostatitis**—is inflammation of the prostate gland that results in urinary frequency and urgency, burning or painful urination (dysuria), and pain in the lower back and genital area, among other symptoms. In some cases, prostatitis is caused by bacterial infection and can be treated with antibiotics. But the more common forms of prostatitis are not associated with any known infecting organism. Antibiotics are often ineffective in treating the nonbacterial forms of prostatitis.
- **Proteinuria**—is the presence of abnormal amounts of protein in the urine. Healthy kidneys take wastes out of the blood but leave in protein. Protein in the urine does not cause a problem by itself. But it may be a sign that your kidneys are not working properly.
- **Renal (kidney) failure**—results when the kidneys are no able to regulate water and chemicals in the body or remove waste products from your blood. *Acute renal failure* (ARF) is the sudden onset of kidney failure. This can be caused by an accident that injures the kidneys, loss of a lot of blood, or some drugs or poisons. ARF may lead to permanent loss of kidney function. However, if the kidneys are not seriously damaged, they may recover. *Chronic kidney disease* (CKD) is the gradual reduction

of kidney function that may lead to permanent kidney failure, or end-stage renal disease (ESRD). You may go several years without knowing you have CKD.

- **Urinary tract infections (UTIs)**—are caused by bacteria in the urinary tract. Women get UTIs more often than men. UTIs are treated with antibiotics. Drinking lots of fluids also helps by flushing out the bacteria.

> ✓   **Important Point:** The name of the UTI depends on its location in the urinary tract. An infection in the bladder is called *cystitis*. If the infection is in one or both of the kidneys, the infection is called *pyelonephritis*. This type of UTI can cause serious damage to the kidneys if it is not adequately treated.

- **Urinary incontinence** (loss of bladder control)—is the involuntary passage of urine. There are many causes and types of incontinence, and many treatment options. Treatments range from simple exercises to surgery. Women are affected by urinary incontinence more often than men.
- **Urinary retention** (bladder-emptying problems)—is a common urological problem with many possible causes. Normally, urination can be initiated voluntarily and the bladder empties completely. Urinary retention is the abnormal holding of urine in the bladder. *Acute urinary retention* is the sudden inability to urinate, causing pain and discomfort. Causes can include an obstruction in the urinary system, stress, or neurologic problems. *Chronic urinary retention* refers to the persistent presence of urine left in the bladder after incomplete emptying. Common causes of chronic urinary retention are bladder muscle failure, nerve damage, or obstructions in the urinary tract. Treatment for urinary retention depends on the cause.

Chapter Review Questions

1. The urinary system is sometimes referred to as the _____ system.
2. The organs of the urinary system are not the only organs that contribute to the excretory system.

   - True
   - False

3. The kidneys are the organs that filter the _____, remove the _____, and _____ the wastes in the urine.
4. The cortex and medulla make up the parenchyma, or functional tissue, of the kidney, which contains over a million _____.
5. The renal pelvis is a large cavity that collects the _____ as it is produced.
6. The internal and external urethral _____ control the flow of urine through the urethra.
7. Urine passes from the nephrons into collecting ducts then into the minor _____.
8. _____ rids the body of wastes through the skin.
9. _____ maintains normal blood pressure.
10. The major task of excretion belongs to the _____ system.

References and Suggested Reading

King, R.M. 2003. *Biology Made Simple*. New York: Made Simple Publishers.

Marieb, E.N., and K. Hoelin. 2006. *Human Anatomy and Physiology*, 7th ed. New York: Benjamin Cummings.

NIH. 2004. *Your Urinary System and How It Works*. NIH Publication No. 04-3195. Accessed August 8, 2006, at http://kidney.niddk.nih.gov/kudiseases/pubs/yoururinary/.

Saladin, K.S. 2006. *Anatomy and Physiology: The Unity of Form and Function*. New York: McGraw-Hill.

Seeley, R.R., T.D. Stephens, and P. Tate. 2004. *Anatomy and Physiology*, 7th ed. New York: McGraw-Hill.

SEER. 2006. Anatomy and Physiology. In *SEER Cancer Training*. Accessed August 9, 2006, at http://training.seer.cancer.gov.

CHAPTER 18

Theories of Human Evolution

Each family, among the animals as well as among the plants, comes from the same origin, and even that all animals are come from but one animal, which, in the succession of eras . . . has produced all the races of animals that now exist.

—George Lewis LeClerc (1707–1788)

A major task for biologists is to explain organic diversity in terms of evolutionary principles: to explain why so many different types of species have arisen, and why they vary so much in form, function, and behavior.

—P.R. Grant (1986)

Topics in This Chapter

- Pre-Darwinian Notions
- Charles Darwin
- The Genetic Basis for Evolutionary Change
- Natural Selection: Modes of Selection
- Speciation
- Patterns of Evolution
- Extinction

The old saying goes: "If you want to maintain peace with those you associate with, do not discuss politics and religion." Most of us learn quickly that it is prudent to avoid these 2 topics, especially in the presence of strangers. Thus, with time and experience we learn to channel our spoken points of view toward less volatile topics, such as weather, sports, or summer vacation trips—all of which are (generally speaking, of course) safer topics to discuss.

There is another topic of discussion—possibly the most radical idea in human history—that can be quite volatile, either from a political and/or religious point of view, which we also learn to avoid discussing with unfamiliar people. This topic, evolution, the subject of this chapter, is certainly one that manifests strong differences in opinion. For example, we would not normally discuss the following points with any stranger on the street (or anywhere else, for that matter):

- All life on Earth sprang forth from some form of biological soup, emanating from some primordial sea.
- Later on, human life evolved from some type of ancient ape-like creature.
- Human life is still evolving.

Indeed, if we want to stir up a hornet's nest, all we need do is insert the above statements into our conversation—not always a prudent move, for sure.

Notwithstanding the potential volatility of discussing evolution—defined as the cumulative changes that occur in a population over time—the topic can't be avoided in any rational discussion of human biology. Simply, evolution is the basis for the study of modern biology. As distinguished biologist Theodosius Dobzhansky once remarked, "Nothing in biology makes sense without the concept of evolution." Having said this, keep in mind that no one alive today can definitively explain evolution. This is the case, of course, because no one has witnessed the entire history of life on Earth. Moreover, no one has observed the process of speciation from start to finish. Thus, the obvious question is: Does this mean we can't study evolution or draw conclusions about the evolutionary process? The simple answer is no. The complex answer is also no. Another pertinent question: Since no one has observed the entire process, how can we determine if evolution has occurred and is still occurring? Answer: Deduction.

Many of the concepts commonly accepted today are based on hypothesis or theory. For example, the phenomena of electricity are well understood—we understand what energizes our lightbulbs and cells phones. There are numerous hypotheses and theories about electricity. The fact remains, however, we are still not able to absolutely define what electricity is. Like electricity, evolution happens, it is real, it continues but we do not know for sure how, why, or when, and in many cases, how—we simply do not have the 411 on evolution or electricity.

In regard to our present view of evolution, the most we can expect is interpretation based on various theories and records. For example, modern understanding of human origins is derived largely from the findings of paleontology, anthropology, and genetics, and involves the process of natural selection. Notwithstanding gaps in the fossil record because of differential preservation that prevents the complete specification of the line of human descent, it is apparent that humans (*H. sapiens*) share clear anatomical, genetic, and historic relationships to other primates. Humans bear particularly close affinity to other members of a group known as hominoids, or apes, which includes orangutans, gibbons, gorillas, chimpanzees, and humans. Hominids (humans and their immediate ancestors) are notable among hominoids for their bipedal locomotion, slow rate of maturation, large brain size, and, at least among recent hominids, the development of a relatively sophisticated capacity for language and social activity.

The bottom line: As Peter Grant (1986) points out (controversial or not), it is the task of biologists to explain diversity in terms of evolutionary principles. That is, biologists must work to explain "why so many different types of species have arisen, and why they vary so much in form, function, and behavior." Moreover, it is important to remember that since evolution is a hypothesis, and if the predicted patterns are not found, then we reject evolution as a hypothesis and come up with something else as a hypothesis. Further, it must be pointed out that, to date, the Darwinian theory of evolution has withstood the test of time and thousands of scientific experiments; nothing has disproved it since Darwin first proposed it more than 150 years ago. A scientific theory stands until proven wrong; it is never proven correct.

DEFINITION OF KEY TERMS

- **Adaptation**—any heritable characteristic of an organism that improves its ability to survive and reproduce in its environment.
- **Adaptive radiation**—the diversification, over evolutionary time, of a species or group of species into several different species or subspecies that are typically adapted to different ecological niches.
- **Bipedal**—walking on 2 legs.
- **Brachiation**—locomotion by swinging arm over arm.
- **Evolution**—change through time, usually with reference to biological species, but may also refer to changes within cultural systems.
- **Foramen magnum**—the hole in the base of the skull through which the spinal cord emerges, around the outside of which the first vertebra articulates.
- **Fossil**—the mineralized bone of an extinct animal; the remains of past life-forms.
- **Gene pool**—the sum of all of the genetic components in a population.
- **Genetics**—the branch of biology that deals with the study of heredity or inheritance.
- **Hominid**—modern human beings and our ancestors, generally defined as the primates who habitually walk erect.
- **Hominoidea**—the taxonomic group or family that includes the human and ape members of the primates, both the fossil and modern forms.
- **Macroevolution**—deals with how organisms evolved, and the major shaping events and timeline.
- **Microevolution**—deals with how genetic changes lead to speciation.
- **Natural selection**—evolutionary changes based on the varying reproductive success of individuals within a species.
- **Population**—a group of interbreeding organisms.
- **Primate**—large-brained, mostly tree-dwelling mammals with 3-dimensional color vision and grasping hands. Humans are primates.
- **Quadrupedal**—walking on all 4 legs.
- **Speciation**—the formation of one or more species from a single species, or the increase of biological diversity.
- **Species**—the fundamental unit of biological classification. A group of individuals that is similar in structure, function, and development, with the potential to interbreed and produce fertile offspring.

Pre-Darwinian Notions

ANCIENT GREEKS

In studying the evolution of the "theory of evolution" that is widely accepted today, we are directed to ancient Greek philosophy, where it made its early appearance. While it is true that the Greeks had no term exactly equivalent to "evolution," they did have some early inkling that pointed to the conception. For example, when Thales (circa 624–546 BC), known as one of the Seven Sages of Greece, asserted that all things originated from water; when Anaximenes (585–525 BC) called air the principle of all things—the source of all that exists, and regarded the subsequent process as a thinning or thickening, they must have considered individual beings and the phenomenal world as a result of evolution, even if they did not develop the concept in detail. Anaximander (circa 610–546 BC), an early proponent of exact science who is often regarded as a precursor of the modern theory of development, "deduced living beings developed gradually from moisture under the influence of warmth, and suggested the view that men originated from animals of another sort, since if they had come into existence as human beings, needing fostering care for a long time, they would not have been able to maintain their existence" (IEP 2006).

In Empedocles' writings (490–430 BC; best known for his philosophy that all matter is made up of 4 elements: water, earth, air, and fire), as with Epicurus (341–270 BC; philosophy positing that good and bad are derived from their associated physical sensations) and Lucretius (circa 94–49 BC; *On the Nature of Things*), "who follow in his footsteps, there are rudimentary suggestions of the Darwinian theory in its broader sense; and here too, as with Darwin, the mechanical principle comes in; the developmental process is adapted to a certain end by a sort of natural selection, without regarding nature as deliberately forming its results for these ends" (IEP 2006).

ARISTOTLE'S *SCALA NATURAE*

Aristotle (384–382 BC) documented what has been referred to as the *scala naturae*, also called the "Great Chain of Being" or "Ladder of Nature." He stated that species are arranged in order of increasing perfection (humans at the top, of course). According to Aristotle, species are fixed and unchanging. Also, the earth is assumed to be quite young (several thousands of years). The *scala naturae* idea began to break down with new discoveries that didn't fit easily into it:

- Microscopic organisms
- New lands, new creatures
- Investigations into geology suggesting a very old Earth
- Fossils of extinct plants and animals

LINNAEUS

From the 1600s through the 1800s, the idea that the "path to God is by contemplating his works" was a romantic and popular one which at least justified and perhaps motivated much early natural history work (Olsen 2001). This philosophy led to attempts at classifications. However, one fundamental problem was—what are the units of life to be classified? Then upon the world's stage entered "God's Registrar," better known as Linnaeus.

In the mid-1700s, the Swedish biologist Carl Linnaeus developed the scientific nomenclature system still used in biology; he took it upon himself to classify the entire natural world. He placed humans in the order of the Primates along with apes and monkeys, but seems to have encountered little criticism.

"Linnaeus recognized that interbreeding showed that there was a natural break between organisms that would freely interbreed and those that would not. This was a fact known well by breeders of animals. Linnaeus saw that there was a "unity of type." He showed that the basic unit of natural classification was not the individual but the "species," and that they were recognizable by the fact that individuals within a species would freely interbreed. That is, a species is held together by sex. Linnaeus' studies showed that the biological definition of a species also holds true for animals" (Olsen 2001).

✓   **Important Point**: Linnaeus' system was purely descriptive, making no claims about origins. Also, its hierarchical nature meshed well with the hierarchical social and political systems of the time.

✓   **Important Point**: "In most cases, species are recognized by the fact that members of a species tend to look much more like each other than they do to members of other species" (Olsen 2001).

"In 1758, Linnaeus published his magnum opus: *Systema Naturae, 10th ed.* In this work he outlined not only the known species of animals and plants, but also what has come to be known as the "Linnaean hierarchy" of taxonomic levels. Linnaeus noticed that while the fundamental unit was the species, species could be grouped by similarity of structure into larger groups: genera, families, orders, classes, phyla, and kingdoms (and others in between). We call any entity within this hierarchy a taxon (plural—taxa). Thus, a species is a taxon as is a family. This is an inclusive hierarchy. That is, kingdoms consist of and contain classes, classes consist of and contain orders, orders consist of and contain families, families consist of and contain genera, and genera consist of and contain species" (Olsen 2001).

✓   **Important Point**: "It is important to note that in the Linnaean hierarchy species are handled specially—a fact the student of biology needs to know and be familiar with" (Olsen 2001).

The Linnaean hierarchy for humans is:

Phylum Chordata
    Class Mammalia
        Order Primates
            Family Hominidae
                Genus *Homo*
                    Species *Homo sapiens*

The fact that this classification system worked became clear very quickly and was universally adopted. One reason for wide acceptance of Linnaeus' classification proposal was that, though he "catalogued species as if invariable, he noted that occasional hybrids between species suggested to him they were not quite so fixed. In fact, although he began by believing that species were fixed entities create by God as is, later in life he began to believe otherwise. Simply, Linnaeus came to believe that species were children of time, capable of transformation in form through time—in other words, species could evolve" (Olsen 2001); evolution is a work in progress.

LAMARCK

Charles Darwin acknowledged Jean-Baptist Lamarck (1744–1829) as a forerunner of evolution. Darwin wrote in 1861:

> Lamarck was the first man whose conclusions on the subject (evolution) excited much attention. This justly celebrated naturalist first published his views in 1801 . . . he first did the eminent service of arousing attention to the probability of all changes in the organic, as well as in the inorganic world, being the result of law, and not of miraculous interposition.

Darwin leveled this praise because Lamarck came up with the first reasonable theory of organic evolution. Lamarck recognized that:

- Organisms change in their environment.
- They discontinuously change and reorganize.

However, Lamarck could not deal with the concept of extinction.

✓   **Important Point**: Lamarck believed that species could change through time by passing on traits acquired during an individual's life to their offspring—the so-called "acquired characteristics."

CUVIER

Georges Cuvier (1769–1832) is probably best known for his realization that extinction is real. This view was a great advance for science. Further, Cuvier's integration of organismal function into the study of form has proved to be a powerful tool for biologists. His views ran right into Lamarck's. However, it can be said that modern evolutionary thought has synthesized both Cuvier's views and those of his opponents into a coherent whole.

✓   **Important Point**: There were other evolutionary concepts around, but like Lamarck's ideas, none took hold because they lacked a clear and observable mechanism—until Darwin, that is.

Charles Darwin

In 1831, Charles Darwin began a 5-year journey on the H.M.S. *Beagle* as a naturalist. During the voyage, he saw many unfamiliar patterns of geography and organismal variation that needed to be explained. Based on his observations and studies, he developed the Theory of Evolution as explained in his book, *On the Origin of Species by Means of Natural Selection*, published in 1859. His theory sets about explaining the progressive changes that occur within species down the generations as well as the formation of new species, when environmental pressures have differential effects on the reproductive success of all individuals.

NATURAL SELECTION

Darwin's theory has 2 aspects to it, namely *natural selection* and *adaptation*, that work together to shape the inheritance of alleles within a given population. **Natural selection** shapes adaptations and differentiates between the reproductive successes of individuals. That is, individuals in a population who are well adapted to a particular set of environmental conditions have an advantage over those who are not so well adapted. The advantage comes in the form of survival and reproductive success. For example, those individuals who are better able to find and use a food resource will, on average, live longer and produce more offspring than those who are less successful at finding food. Inherited traits that increase individuals' fitness are then passed to their offspring, thus giving the offspring the same advantages.

Probably one of the best illustrations of Darwin's natural selection and adaptation in progress is demonstrated by the observations and studies of Peter R. and Rosemary Grant, documented in Peter R. Grant's 1986 epic work, *Ecology and Evolution of Darwin's Finches*. The Grants spent years observing, tagging, and measuring Galapagos finches (aka Darwin's finches) and their environment. During that time, they documented environmental changes and how these changes favored certain individuals within the population. Those individuals survived and passed their characteristics on to the next generation.

During their observations, on the 100-acre Daphne Major Island (their natural laboratory), the Grants watched the struggle for survival among individuals in 2 species of small birds, called Darwin's finches. The struggle is mainly about food—different types of seeds—and the availability of that food is dramatically influenced by year-to-year weather changes.

The Grants wanted to find out whether they could actually see the force of natural selection at work, judging by which birds survived the changing environment. For the finches, body size and the size and shape of their beaks are traits that vary in adapting to environmental niches or changes in those niches. Body and beak variation occur randomly. The birds with the best-suited bodies and beaks for the particular environment survive and pass along the successful adaptation from one generation to another through natural selection.

During a drought in 1977, natural selection winnowed certain finches harshly. That year, the vegetation withered. Seeds of all kinds were scarce. The small, soft ones were quickly exhausted by the birds, leaving mainly large, tough seeds that the finches normally ignore. Under these drastically changing conditions, the struggle to survive favored the larger birds with deep, strong beaks for opening the hard seeds.

Smaller finches with less-powerful beaks perished.

So the birds that were the winners in the game of natural selection lived to reproduce. It just so happened that the big-beaked finches were the ones favored by the particular set of conditions Nature imposed that year.

The next step? Evolution. The Grants found that the offspring of the birds that survived the 1977 drought tended to be larger, with bigger beaks. So the adaptation to a changed environment led to a larger-beaked finch population in the following generation.

As the Grants later found, adaptation can go either way, of course. Unusually, rainy weather in 1984–1985 resulted in more small, soft seeds on the menu and fewer of the large, tough ones. Consequently, the birds best adapted to eat those seeds because of their smaller beaks were the ones that survived and produced the most offspring.

Evolution had cycled back the other direction—Charles Darwin would be proud.

There are 8 lines of evidence which led Darwin to propose his theory of natural selection:

1. South America is the home of unique animals not found on other continents, and extinct giant relatives of the living forms are found there as fossils (Levin 2006).
2. Species of marine organisms on either side of Panama (in the Atlantic and Pacific Oceans) are very different, although they are only a few miles apart (barrier to dispersal).
3. Tortoises of the Galapagos Islands are different on each island. Suggests a common ancestry. They differentiated in form as a result of living apart in different environments.
4. Finches in the Galapagos Islands are adapted to exploit a variety of different niches. Beaks show a lot of variation and specialization for different functions. Similar types of changes seen in the Hawaiian honeycreepers (Levin 2006).
5. Similarity of embryos of all vertebrates suggests a common ancestry.
6. Vestigial organs suggest a common ancestry (serve no apparent purpose, but resemble functioning organs in other animals). Whales have useless pelvic bones (and

occasionally rear feet) resembling those in other mammals. Why develop a useless appendage or structure?

7. Homologous organs and bone configurations have a common ancestry (toes of land-dwelling mammals vs. bat wings). Evidence of this abounds in animal and plant kingdoms.
8. Changes in domestic animals through selective breeding.

These lines of evidence and deductive reasoning led Darwin to derive 5 parts to natural selection:

1. Heredity of most features.
2. Heritable variation in the population.
3. Variation leads to differential rates in survival and reproductive success among the variations; not all young produced in each generation can survive.
4. Differential survival and reproduction leads to a shift in the frequency of characters, leading to a shifting of mode within the species.
5. If this process goes on long enough, parent and daughter species can no longer interbreed.

Adaptations are anatomical structures, physiological processes, or pattern of behavior that contribute to ancestral survival through the unique suitability of those traits/characteristics (Crawford 1998).

✓   **Important Point**: Populations and species evolve, individuals do not.

The Genetic Basis for Evolutionary Change

As mentioned, species formation usually requires geographic isolation to bring about reproductive isolation. Also, natural selection results in adaptation to local environmental conditions. In this section, the links between genetics and evolution are explored.

As previously discussed, important advances in genetics and cell biology established that hereditary units called genes behave according to certain principles first discovered by Mendel, and that genes are located on chromosomes. During cell division leading to the formation of sex cells, or gametes, cells divide and chromosomes separate according to the process of meiosis, which explains the segregation and recombination of traits that Mendel described.

In order to understand the genetic basis for evolutionary change, it is crucial to answer 2 questions:

1. Where does genetic variation come from?
2. What are the causes of changes in the genetic composition of a population over time?

SOURCES OF GENETIC VARIATION

Genetic variation is important because it provides the opportunity for evolutionary change. If it results in individuals who have a greater chance of surviving and reproducing, natural selection will cause those individuals to leave more offspring, and evolution will take place. The availability of new genetic material is one of the primary determinants of the rate of evolution.

Four processes are of paramount importance in causing change in the genetic material:

- **Mutational changes in the DNA**—results in the creation of new genetic material
- **Transposition of the position of genes**—results in creation of new genetic combinations
- **Recombination by crossing-over during meiosis**—results in creation of new genetic combinations
- **Meiosis and sex**—creation of new genetic combinations

HARDY-WEINBERG PRINCIPLE

The basic principle of the genetics of populations is called the **Hardy-Weinberg Principle**. Essentially, the principle says that: *In a large population, with random mating, and in the absence of forces that would change their proportions, the proportion of alleles at a given locus will remain constant.* Stated differently, in the absence of selection, mutation, migration, or random chance, nonrandom mating genotype frequencies can be predicted from allele frequencies and allele frequencies in a population will not change from one generation to the next. This principle was important because it showed that random mating alone does not cause evolution.

✓ **Important Point**: Although random mating alone does not cause evolution, nonrandom mating (also called *assertive mating*) does. This limits breeding to members of a population that possess certain desirable traits, such as the strongest male lion or the male bird with the red crest.

Natural Selection: Modes of Selection

There are 3 general modes of selection:

- **Directional selection**—occurs when selection favors one extreme trait value over the other extreme; that is, natural selection pushes gene frequencies toward 1 of 2 extremes, away from the average character trait. This typically results in a change in the mean value of the trait under selection. An example is evolution in horse limb morphology illustrating directional selection—over time, natural selection favored individuals with limbs adapted for running on open grassland areas.

- **Disruptive selection**—occurs when selection favor the extreme trait values over the intermediate trait values (i.e., diversifying selection favors 2 or more extreme forms of a trait). In this case, the variance increases as the population is divided into 2 distinct groups. Disruptive selection plays an important role in speciation. An example is the artic charr (fish) 4 four morphs illustrate disruptive selection: Natural selection has produced 4 different body shapes, color patterns, and sizes in this fish.
- **Stabilizing selection**—occurs when selection favors the intermediate trait value over the extreme values; that is, natural selection maintains population gene frequencies clustered around an average. Populations under this type of selection typically experience a decrease in the amount of additive genetic variation for the trait under selection. An example is the chambered nautilus, a life-form which has remained relatively unchanged in form for the last 400 million years or so (i.e., the nautilus body form has stabilized to an average shape and size).

Speciation

Speciation is the formation of 1 or more species from a single species, or the increase of biological diversity. Many biologists believe that it is the key to understanding evolution. Types of speciation are defined by the way in which populations become isolated. There are 3 basic types:

- **Allopatric speciation** is the most common form of speciation seen today. It occurs when populations of a species become geographically isolated.
- **Parapatric speciation** is extremely rare. It occurs when populations are separated by an extreme change in habitat rather than a physical barrier.
- **Sympatric speciation** occurs when populations of a species that share the same habitat become reproductively isolated from each other.

Patterns of Evolution

Natural selection can lead to the formation of new species. Sometimes many species evolve from a single ancestral species. Similarities in skeletal and muscular structure of Hawaiian honeycreepers led scientists to conclude that the 23 species of honeycreepers evolved from 1 ancestral species. Such an evolutionary pattern, in which many related species evolved from a single ancestral species, is called **adaptive radiation**. Thus, adaptive radiation (aka divergent evolution) is the evolving of species from a common ancestor, radiating out like the spokes of a wheel. An example is the mammal population after the dinosaurs died. In **convergent evolution**, on the other hand, unrelated organisms look similar because they evolved in similar environments. **Coevolution** is the joint change of 2 or more species in close interaction. That is, it occurs when species closely connected to one another evolve together. One example of coevolution can be seen between plants and the animals that pollinate them.

✓   **Important Point**: Patterns of evolution are examples of how the diversity of life on Earth is due to the ever-changing interaction between a species and its environment.

Extinction

Extinction is an on-going process.
    Extinction is a work in progress.
    Extinction happens all the time.
    The fact is, scientists estimate that 99% of all species that ever lived are now extinct. It can be said that if the world has fairly even diversity through time, existing species will become extinct about as often as new species evolve. The bottom line: Extinction is the expected fate of species, rather than a rarity.
    The preceding statements may be eye-openers to those untrained in or unaware of the phenomena of taxa extinction. When the average person thinks about extinction (that is, if he or she ever does think about it), probably the first thought that comes to mind is the extinction of the dinosaurs. From the time we are children we are exposed to T-Rex and his/her peers of that ancient time. This exposure first comes from pictures, television, and toy dinosaurs. Indeed, many of us go through life intrigued with dinosaurs—and with their ultimate fate. As children and after our first exposure to the dinosaurs it is not long before we all ask(ed): What happened to the dinosaurs? There are numerous theories about the causes of the dinosaurs' demise. With time and education, we are all exposed to these differing theories. Many of us come to believe that mass extinction—a global phenomenon as in the case of the total destruction of the dinosaurs and other species of that time—was caused not caused by some ongoing, definitively identifiable "extinction bullet." Instead, because we do not know what we do not know about ancient times and causal factors related to past mass extinctions, extinction was likely the result of a few plausible extinction agents:

- An enormous volcanic eruption
- An extraterrestrial impact by an asteroid
- A rapid change in sea level affecting global ecology and climate
- Massive changes in global climate and atmosphere
- Variation in solar radiation

    It should be pointed out, however, that many experts believe that past mass extinctions were not the result of a single factor but the result of multiple factors (multiple "extinction bullets"); that is, mass extinction was caused by factors including, volcanoes, comets, variation in solar radiation, shifting of continents, sea level changes, etc. Scientists estimate that there are 5 to 30 million different species on Earth today, and it would not be surprising (to the author, at least) if there were that range of opinions on the causal factors of past mass extinctions.

While there is dispute and several varying theories as to the causal factors of extinction events, there is no doubt in the scientific community that extinction events have indeed occurred. The scientific record clearly demonstrates that in that past at least 5 mass extinction events have occurred. These mass extinctions were responsible for eradication of entire species; destruction of entire ecological systems; and the collapse of critical food webs.

For some who have read the above paragraph while sitting in the comfort of their homes the question(s) may be: So what? What does all this mean? What effect does this have on me? If mass extinctions did actually occur, who cares?

These questions are difficult to answer and beyond the scope of this text. However, there are a few salient points to be made. The first point is that mass extinctions leave habitat (niches) open, affording survivors multiple opportunities. Probably the second point was best stated by Jablonski (2001, 31):

> To the conservation biologist, there is little positive to be said about extinction. From an evolutionary perspective, however, extinction is a double-edged sword. By definition, extinction terminates lineages and thus removes unique genetic variation and adaptations. But over geological time scales, it can reshape the evolutionary landscape in more creative ways, via the differential survivorship of lineages and the evolutionary opportunities afforded by the demise of dominant groups and the postextinction sorting of survivors. The interplay between the destructive and generative aspects of extinction, and the very different time scales over which they appear to operate, remains a crucial but poorly understood component of the evolutionary process.

Chapter Review Questions

1. _____ is any heritable characteristic of an organism that improves its ability to survive and reproduce in its environment.
2. _____ is the diversification of a species or group of species into several different species or subspecies.
3. Deals with how genetic changes lead to speciation: _____.
4. _____ is the formation of one or more species from a single species.
5. The _____ stated that species are arranged in order of increasing perfection.
6. _____ recognized that interbreeding showed that there was a natural break between organisms that would freely interbreed and those that would not.
7. Darwin acknowledged _____ as a forerunner of evolution.
8. Best known for his realization that extinction is real: _____.
9. _____ shares adaptations and differentiates between the reproductive successes of individuals.
10. The basic principle of the genetics of populations is called the _____.

References and Recommended Reading

Adams, A.B. 1969. *Eternal Quest: The Story of the Great Naturalists.* New York: G.P. Putnam's Sons.

Buffon, G.L.L. 1835. *Oeuvres complètes de Buffon. Nouvelle Edition. Tome Premier.* Paris: E.L.C. Mauprivez.

Crawford, C. 1998. *The Theory of Evolution in the Study of Human Behavior: An Introduction and Overview.* In C. Crawford and D.L. Krebs (Eds.), *Handbook of Evolutionary Psychology: Ideas, Issues, and Applications,* 3–42. Mahwah, NJ: Lawrence Associates, Inc.

Darwin, C. 1859. *On the Origin of Species.* London: John Murray.

Darwin, C. 1861. "An Historical Sketch of the Progress of Opinion on the Origin of Species, Previously to the Publication of This Work." Appended to 3rd and later editions of *On the Origin of Species.*

Eiseley, L. 1958. *Darwin's Century: Evolution and the Men Who Discovered It.* New York: Doubleday.

Farley, K.A., et al. 1998. Geochemical Evidence for a Comet Shower in the Late Eocene. *Science* 180: 1250–1253.

Grant, P.R. 1986. *Ecology and Evolution of Darwin's Finches.* Princeton, NJ: Princeton University Press.

Hall, B.K. 1992. *Evolutionary Developmental Biology.* London: Chapman and Hall.

IEP. 2006. Evolution. *The Internet Encyclopedia of Philosophy.* Accessed July 7, 2006, at www.iep.utm.edu/e/evolutio.htm.

Jablonski, D. 2001. Lessons from the Past: Evolutionary Impacts of Mass Extinctions. *Proceedings of the National Academy of Sciences of the United States of America* 98: no. 10.

Johanson, D., and B. Edgar. 1996. *From Lucy to Language.* London: The Orion Publishing Group.

LeClerc, G-L. 1749. *Historie naturelle.* Paris: L'Homme.

Levin, H.L. 2006. *The Earth through Time.* New York: John Wiley and Sons.

Lewontin, R. 1974. *The Genetic Basis of Evolutionary Change.* New York: Columbia University Press.

McKinney, M.L. 1997. Extinction, Vulnerability and Selectivity: Combining Ecological and Paleontological Views. *Annual Review of Ecology and Systematics* 28: 495–516.

Olsen, P. 2001. *Dinosaurs and the History of Life.* Accessed June 28, 2006, at http://rainbow.ldeo.columbia.edu.

Packard, A.S. 1901. *Lamarck, the Founder of Evolution: His Life and Work.* New York: Longmans, Green, and Co.

Quinlan, G.D. 1993. Plant X: A Myth Exposed. *Nature* 363: 18–19.

Rudwick, J.J.S. 1985. *The Meaning of Fossils: Episodes in the History of Palaeontology.* Chicago: University of Chicago Press.

Smith, J.C. 1993. *Georges Cuvier: An Annotated Bibliography of His Published Works.* Washington, DC: Smithsonian Institution Press.

Wessells, N.K., and J.L. Hopson. 1988. *Biology.* New York: Random House.

Part III

INTRODUCTION
TO ECOLOGY

CHAPTER 19

Introduction to Ecology

We poison the caddis flies in a stream and the salmon runs dwindle and die. We poison the gnats in a lake and the poison travels from link to link of the food chain and soon the birds of the lake margins become victims. We spray our elms and the following springs are silent of robin song, not because we sprayed the robins directly but because the poison traveled, step by step, through the now familiar elm-leaf-earthworm-robin cycle. These are matters of record, observable, part of the visible world around us. They reflect the web of life—or death—that scientists know as ecology.

—Rachel Carson (1962)

Topics in This Chapter

- Levels of Organization
- Biosphere
- Ecosystem
- Biomes
- Community Ecology
- Population Ecology
- Flow of Energy in an Ecosystem

As Rachel Carson points out, what we do to any part of our environment has an impact upon other parts. In other words, there is an interrelationship between the parts that make up our environment. Probably the best way to state this interrelationship is

---

Note: Much of the material presented in this chapter originated from the author's copyrighted 1995 dissertation on stream ecology and self-purification. The material was subsequently published in various texts and formats including Government Institutes 1999/2006 editions of *Environmental Science and Technology: Concepts and Applications*, 1st/2nd eds.

to define ecology. "Ecology is the science that deals with the specific interactions that exist between organisms and their living and nonliving environment" (Tomera 1989). The term "ecology" is derived from the Greek *oikos*, meaning home. Ecology, then, means the study of organisms at home—in other words, the study of an organism at its home. Ecology is the study of the relation of an organism or a group of organisms to their environment. In a broader sense, ecology is the study of the relation of organisms or groups to their environment.

✓   **Important Point**: No ecosystem can be studied in isolation. If we were to describe ourselves, our histories, and what made us the way we are, we could not leave the world around us out of our description.

Charles Darwin explained ecology in a famous passage in the *Origin of Species*, a passage that helped establish the science of ecology. A "web of complex relations" binds all living things in any region, Darwin writes. Adding or subtracting even a single species causes waves of change that race through the web, "onwards in ever-increasing circles of complexity." The simple act of adding cats to an English village would reduce the number of field mice. Killing mice would benefit the bumblebees, whose nest and honeycombs the mice often devour. Increasing the number of bumblebees would benefit the heartsease and red clover, which are fertilized almost exclusively by bumblebees. So adding cats to the village could end by adding flowers. For Darwin, the whole of the Galapagos archipelago argues this fundamental lesson. The volcanoes are much more diverse in their ecology than their biology. The contrast suggests that in the struggle for existence, species are shaped at least as much by the local flora and fauna as by the local soil and climate. "Why else would the plants and animals differ radically among islands that have the same geological nature, the same height, and climate" (Darwin 1998, 112)?

The environment includes everything important to the organism in its surroundings. The organism's environment can be divided into 4 parts:

1. **Habitat and distribution**—its place to live
2. **Other organisms**—whether friendly or hostile
3. **Food**
4. **Weather**—light, moisture, temperature, soil, etc.

There are 2 major subdivisions of ecology: autecology and synecology. Autecology is the study of the individual organism or a species. It emphasizes life history, adaptations, and behavior. It is the study of communities, ecosystems, and biosphere. Synecology is the study of groups of organisms associated together as a unit (Odum 1971).

An example of autecology would be when biologists spend their entire lifetime studying the ecology of the salmon. Synecology, on the other hand, deals with the environmental problems caused by mankind. For example, the effects of discharging phosphorous-laden effluent into a stream involve several organisms. The activities of human beings have become a major component of many natural areas. As a result, it is important to realize that the study of ecology must involve people.

DEFINITION OF KEY TERMS

Each division of ecology has its own set of terms that are essential for communication between ecologists and those studying ecology. Basic ecological terms are defined in the following:

- **Abiotic factor**—the nonliving part of the environment composed of sunlight, soil, mineral elements, moisture, temperature, topography, minerals, humidity, tide, wave action, wind, and elevation.

  ✔ **Important Note**: Every community is influenced by a particular set of abiotic factors. While it is true that the abiotic facts affect the community members, it is also true that the living (biotic factors) may influence the abiotic factors. For example, the amount of water lost through the leaves of plants may add to the moisture content of the air. Also, the foliage of a forest reduces the amount of sunlight that penetrates the lower regions of the forest. The air temperature is therefore much lower than in nonshaded areas (Tomera 1989).

- **Aquatic**—living or growing in water.
- **Autotroph (primary producer)**—any green plant that fixes energy from sunlight to manufacture food from inorganic substances.
- **Bacteria**—among the most common microorganisms in water, bacteria are primitive, single-celled microorganisms (largely responsible for decay and decomposition of organic matter) with a variety of shapes and nutritional needs.
- **Biochemical oxygen demand (BOD)**—a widely used parameter of organic pollution applied to both wastewater and surface water, involving the measurement of the dissolved oxygen used by microorganisms in the biochemical oxidation of organic matter.
- **Biogeochemical cycles**—cyclic mechanisms in all ecosystems by which biotic and abiotic materials are constantly exchanged.
- **Biome**—large-scale communities classified by a main vegetation type and distinctive combinations of plants and animals.
- **Biosphere**—regions of Earth's crust, water, and atmosphere inhabited by living things.
- **Biotic factor (community)**—the living part of the environment composed of organisms that share the same area, are mutually sustaining, interdependent, and constantly fixing, utilizing, and dissipating energy.
- **Climax community**—the terminal stage of ecological succession in an area.
- **Competition**—a critical factor for organisms in any community; animals and plants must compete successfully in the community to stay alive.
- **Community**—in an ecological sense, all the populations occupying a given area.

  ✔ **Important Note**: Leopold (1970, 36) points out: "That land is a community is the basic concept of ecology, but that land is to be loved and respected is an extension of ethics. That land yields a cultural harvest is a fact long known, but latterly often forgotten."

- **Decomposition**—the breakdown of complex material into simpler substances by chemical or biological processes.
- **Dissolved oxygen (DO)**—the amount of oxygen dissolved in a stream is an indication of the degree of health of the stream and its ability to support a balanced aquatic ecosystem.
- **Drainage basin (catchment area, stream's watershed)**—the land area that a stream drains.
- **Ecosystem**—the community and the nonliving environment functioning together as an ecological system.
- **Emigration**—the departure of organisms from one place to take up residence in another area.
- **Environment**—everything that is important to an organism in its surroundings. The abiotic factor is composed of sunlight, soil, mineral elements, temperature, moisture, and topography. The biotic community is the natural combination of organisms (plants and animals) that share the same area, are mutually sustaining, interdependent, and constantly fixing, utilizing and dissipating energy.
- **Eutrophication**—the natural aging of a lake or land-locked body of water which results in organic material being produced in abundance due to a ready supply of nutrients accumulated over the years.
- **Foraging**—searching for food.
- **Fungi**—a group of organisms that lack chlorophyll and obtain nutrients from dead or living organic matter.
- **Habitat**—in ecological terms, the place where an organism lives.
- **Heterotroph (living organisms)**—any living organism that obtains energy by consuming organic substances produced by other organisms.
- **Homeostasis**—a natural occurrence during which an individual population, or an entire ecosystem, regulates itself against negative factors and maintains an overall stable condition.
- **Immigration**—the movement of organisms into a new area of residence.
- **Invertebrates**—organisms without a backbone.
- **Limiting factor**—a necessary material that is in short supply and because of the lack of it an organism cannot reach its full potential.
- **Microhabitat**—describes very local habitats.
- **Niche**—the role that an organism plays in its natural ecosystem, including its activities, resource use, and interaction with other organisms.

> ✓   **Important Note:** Each species fills a number of different niches that together make its microhabitat and lifestyle.

- **Nonpoint pollution**—sources of pollutants in the landscape, such as agricultural runoff.
- **Nitrogen fixation**—ability of organisms to take nitrogen gas out of the air and make it into biologically useful nitrogen.
- **Nutrient**—a material that serves as food or provides nourishment.
- **Organic**—derived from living organisms.

- **Point source**—source of pollutants that involves discharge of pollutants from an identifiable point, such as a smokestack or sewage treatment plant.
- **Pollution**—an adverse alteration to the environment by a pollutant.
- **Population**—a group of organisms of a single species that inhabit a certain region at a particular time.
- **Runoff**—after an organic waste has been applied to a soil, the possibility exists that some of this waste may be transmitted by rainfall, snowmelt, or irrigation runoff into surface waters.
- **Scavengers**—animals that feed on dead or decaying organic matter.
- **Species**—the basic category of biological classification consisting of similar organisms that are capable of mating and reproduction.
- **Succession**—a process that occurs subsequent to disturbance, and that involves the progressive replacement of biotic communities with others over time.
- **Trophic level**—the feeding position occupied by a given organism in a food chain, measured by the number of steps removed from the producers.
- **Water cycle**—biogeochemical cycle that moves and recycles water in various forms through the biosphere.

Levels of Organization

"The best way to delimit modern ecology is to consider the concept of levels of organization" (Odum 1983, 3). Levels of organization can be simplified as follows:

Species → Population → Communities → Ecosystem → Biome → Biosphere

In this relationship, organs form an organism; organisms of a particular species form a population; populations occupying a particular area form a community; communities, interacting with nonliving or abiotic factors, separate in a natural unit to create a stable system known as the ecosystem (the major ecological unit); and the part of Earth in which an ecosystem operates is known as the biosphere. Tomera (1989) points out that "every community is influenced by a particular set of abiotic factors." The abiotic part of the ecosystem is represented by inorganic substances such as oxygen, carbon dioxide, several other inorganic substances, and some organic substances.

The physical and biological environment in which an organism lives is referred to as its habitat. Within each level of organization of a particular habitat, each organism has a special role. The role the organism plays in the environment is referred to as its niche. A niche might be that the organism is food for some other organism or is a predator of other organisms. Odum (1975) refers to an organism's niche as its "profession." In other words, each organism has a job or role to fulfill in its environment. Although 2 different species might occupy the same habitat, "niche separation based on food habits" differentiates between 2 species (Odum 1983).

Important Note: In order for an ecosystem to exist, a dynamic balance must be maintained among all biotic and abiotic factors—a concept known as *homeostasis*.

Biosphere

The term *biosphere* (coined by Russian scientist Vladimir Vernadsky in 1929) is the entire portion of the earth inhabited by all living organisms, including humans, and all organic matter that has not yet decomposed. The components of the biosphere are the hydrosphere, the lithosphere and the atmosphere.

- **Hydrosphere**—the biosphere as we know it would not exist without water; it is essential for all living organisms on Earth.
- **Lithosphere**—the lithosphere and biosphere are intimately connected through soils, which consist of a mixture of air, organic matter, mineral matter, and water.
- **Atmosphere**—is the gaseous envelope that surrounds the earth and provides the oxygen and other gases essential to life as we know it on Earth.

Ecosystem

Ecosystem is a term introduced by Tansley to denote an area that includes all organisms therein and their physical environment. The ecosystem is the major ecological unit in nature. "There is a constant interchange of the most various kinds within each system, not only between the organisms but between the organic and the inorganic" (Tansley 1935). Living organisms and their nonliving environment are inseparably interrelated and interact upon each other to create a self-regulating and self-maintaining system. To create a self-regulating and self-maintaining system, ecosystems are homeostatic, meaning they resist any change through natural controls. These natural controls are important in ecology. This is especially the case since it is people, through their complex activities, who tend to disrupt natural controls.

As stated earlier, the ecosystem encompasses both the living and nonliving factors in a particular environment. The living or biotic part of the ecosystem is formed by 2 components: autotrophic and heterotrophic. The autotrophic (self-nourishing) component does not require food from its environment but can manufacture food from inorganic substances. For example, some autotrophic components (plants) manufacture needed energy through photosynthesis. Heterotrophic components, on the other hand, depend upon autotrophic components for food.

 Important Point: The nonliving or abiotic part of the ecosystem is formed by 3 components: inorganic substances, organic compounds (link biotic and abiotic parts), and climate regime.

An ecosystem is a cyclic mechanism in which biotic and abiotic materials are constantly exchanged through biogeochemical cycles. Biogeochemical cycles are defined as follows: *bio* refers to living organisms and *geo* to water, air, rocks, or solids. *Chemical* is concerned with the chemical composition of the earth. Biogeochemical cycles are driven by energy, directly or indirectly from the sun.

In an ecosystem, biotic and abiotic materials are constantly exchanged. Producers construct organic substances through photosynthesis and chemosynthesis. Consumers and decomposers use organic matter as their food and convert it into abiotic components. That is, they dissipate energy fixed by producers through food chains. The abiotic part of a pond, for example, is formed of dissolved inorganic and organic compounds and sediments such as carbon, oxygen, nitrogen, sulfur, calcium, hydrogen, and humic acids. The biotic part is represented by producers such as rooted plants and phytoplanktons. Fish, crustaceans, and insect larvae make up the consumers. Detrivores, which feed on organic detritus, are represented by mayfly nymphs. Decomposers make up the final biotic part. They include aquatic bacteria and fungi, which are distributed throughout the pond.

✓　　**Important Point**: Again, an ecosystem is a cyclic mechanism. From a functional viewpoint, an ecosystem can be analyzed in terms of several factors. The factors important in this study include: biogeochemical cycles, energy, and food chains.

Biomes

When it comes to the terms ecosystem and biome, there is some confusion. What's the difference? The difference between the 2 words is slight. An ecosystem is much smaller than a biome. Actually, the larger biome can be considered as many similar ecosystems throughout the world grouped together. An ecosystem can be as large as the Amazon River ecosystem, or as small as a vernal pool. Thus, a biome is a larger region than an ecosystem and is characterized by habitat conditions and by its community structure. The predominant type of plants that grow there characterizes each biome. In biomes there is a strong relationship between climate and life, which suggests that if we know the climate of an area, we can predict what biome will be found there. The distribution of biomes results from interaction of physical features of the earth. The 2 key physical features of the earth are the amount of solar heat and global atmospheric circulation. Together these factors dictate local climate. The 2 most important climatic factors are precipitation and temperature.

✓　　**Important Point**: Each biome contains many ecosystems whose communities have adapted to differences in climate, soil, and other environmental factors.

The major terrestrial biomes of the world include:

- desert
- grassland
- savanna
- Mediterranean shrublands (chaparral)
- tropical dry forest
- tropical rainforest
- temperate deciduous forest
- taiga or boreal forest
- tundra

DESERTS

Approximately one-third of the earth's land surface is desert, arid land with meager rainfall (<25 cm/yr) that supports only sparse vegetation (depends on water conservation) and a limited population of people and animals. Deserts can be cool or even cold during parts of the year; temperature can vary greatly during a 24-hour period. Desert organisms have evolved adaptations to help them survive: restricting activity to times of year when water is present; avoiding high temperatures by living in deep, cool, and (sometimes) moist burrows; emerging only at night when temperatures are lower; and drinking large quantities of water when it is available (e.g., camels) and then survive long, dry periods. The world's great deserts are located in interiors of continents: Sahara in Africa and Gobi in Asia.

GRASSLANDS

Grasslands, also known as temperate grasslands, prairies, and steppes, develop wherever rainfall is not heavy enough to produce a forest nor light enough to form a desert. Grasslands are widely distributed throughout temperate regions; halfway between the equator and poles. Precipitation is approximately 25–75 cm/year. Generally, grasslands tend to be windy, with hot summers and cold to mild winters. Grasses make up 60%–90% of vegetation—trees are rare except along water courses due to their need for greater amounts of water. Fire in grasslands is important in preventing invasion of trees and in releasing nutrients from dead plants into the soil, contributing to high fertility of grassland soils. Grasslands are often highly productive when converted to agricultural use. The prairies of U.S. and Canada were originally occupied by grasslands. Roots of perennial grasses characteristically penetrate deep into the soil; these soils, therefore, tend to be deep and fertile. Temperate grasslands were often populated by huge herds of grazing mammals. For example, in the U.S. countless numbers of bison and pronghorns once inhabited prairies—these herds are almost gone now and most prairies have been converted into the richest agricultural region on Earth.

SAVANNA

The term *savanna* is believed to have originally come from an Amerindian word, described by Spanish historian Oviedo y Valdes (1535) as "land which is without trees but with much grass either tall or short." By the late 1800s it was used to mean "land with both grass and trees." It now refers to land with grass and either scattered trees, or an open canopy of trees. Precipitation is 50–150 cm/yr but occurs seasonally with periods of heavy rainfall followed by prolonged drought, resulting in a seasonally structured ecosystem. Predominant plants are grasses with widely spaced, drought-resistant trees. Many animals and plants are active only during rainy season. Fire is a common but trees tend to be fire resistant. Savannas are increasingly being converted to agricultural use, causing inhabitants to struggle to survive.

MEDITERRANEAN SHRUBLANDS (CHAPARRAL)

The Mediterranean shrublands biome, also known as chaparral, is found in small sections of most of the continents—the west coast of the United States, the west coast of South America, the Cape Town area of South Africa, the western tip of Australia and the coastal areas of the Mediterranean. Precipitation is 40–100 cm/yr with wet, cool winters and hot, dry summers. Vegetation is dominated by woody shrubs adapted to hot, dry summers. Fire is a common feature.

TROPICAL DRY FOREST

The tropical dry forest biome typically experiences an annual hard dry season. Precipitation range is 50–20 cm/yr and many areas exhibit monsoon climate—several months of heavy rainfall followed by dry periods. Because of highly seasonal rainfall, plants must be drought resistant. Many of the tree species in dry forest are also deciduous. During the driest months these species will drop their leaves. This dry season leaf-drop reduces the water needs of the plant, as there is no evapotranspiration through the leaves.

TROPICAL RAINFOREST

Tropical rainforests are warm (relatively constant) with no frost and very wet (precipitation is >200cm/yr). Rainforests, located near the equator, contain many tall trees. They are made up of the most diverse ecosystems on Earth, containing approximately half of all species of terrestrial plants and animals. For example, in 1 square mile of tropical forest in Brazil there are 1,200 species of butterflies. Rainforest communities are diverse. Each kind of organism is often represented in a given area by only few individuals. Typically, individuals of same species are separated by 1 kilometer or more.

Most nutrients are tied up in biomass and not in the soil; thus, rainforests do not make good farmlands.

TEMPERATE DECIDUOUS FORESTS

The temperate zones are characterized by having 4 seasons and roughly equal-length summers and winters. A deciduous forest is one composed primarily of deciduous trees—those that lose their leaves once a year. Trees are major producers. A temperate deciduous forest is one that is in a temperate zone. Precipitation ranges from 75–100 cm/yr and is evenly distributed. Temperate zones have warm summers and mild winters. Plants grow actively for approximately half the year. These zones are often populated by deer, beaver, bear, and raccoon. Generally, they have a lower number of species but a higher number of individuals per species.

TAIGA

The taiga or boreal forest, one of the largest ecosystems on Earth, exists as a nearly continuous belt of coniferous trees (spruce, fir, larch, etc.) across North America and Eurasia. Coniferous trees with their needle-shaped leaves limit moisture loss. Their pyramid shape accommodates snowfall. Mean annual rainfall is 25–100cm/yr. Climate is humid due to low evaporation resulting from generally low temperatures. Winters are long and cold—soil freezes in winter. Few people live within these regions because of very short growing season for farming.

TUNDRA

The tundra is a treeless area between the ice cap and the tree line of Arctic regions. It is an area having permanently frozen subsoil (permafrost) and supporting low-growing vegetation such as lichens, mosses, and stunted shrubs. Annual precipitation is approximately 25 cm. Spring and summer temperatures are usually less that 10°C (50°F). Tundra is populated by large grazing mammals such as musk-oxen, caribou, reindeer and carnivores such as wolves, foxes, and lynx.

Community Ecology

In ecology, **community ecology** deals with a group of interacting populations in time and space. Communities can be recognized and studied at any number of levels, scales, and sizes. Sometimes a particular subgroup may be specified, such as the fish community in a lake or the soil arthropod community in a forest. Modern community ecology examines patterns such as variation in species, richness, equitability, productivity, and food web structure.

In terms of trophic (feeding) relationships, species comprising communities "function" as:

- **photosynthesizers**—producers
- **herbivores**—primary consumers
- **carnivores**—secondary, tertiary consumers
- **decomposers**
- **omnivores**—obtain food from more than one trophic level

Emergent properties of communities include:

- species diversity
- limits of similarity of competing species
- food web structure
- community biomass and productivity

✓ **Important Point**: A central goal in the field of community ecology is to understand the processes that explain the structure (pattern) of a community; that is, the composition of a species and their abundances and distributions.

Population Ecology

Let's begin with the basics.

POPULATION

Webster's Third New International Dictionary defines population as:

- The total number or amount of things, especially within a given area.
- The organisms inhabiting a particular area or biotype.
- A group of interbreeding biotypes that represents the level of organization at which speciation begins.

POPULATION SYSTEM

Population system or life-system is a population with its effective environment (Berryman 1981).

MAJOR COMPONENTS OF A POPULATION SYSTEM

1. **Population itself**—organisms in the population can be subdivided into groups according to their age, sage, sex, and other characteristics.
2. **Resources**—food, shelters, nesting places, space, etc.

3. **Enemies**—predators, parasites, pathogens, etc.
4. **Environment**—air (water, soil) temperature composition, variability of these characteristics in time and space (Sharov 1997).

Population ecology is the branch of ecology that studies the structure and dynamics of populations. The term "population" is interpreted differently in various sciences. For example, in human demography a population is a set of humans in a given area. In genetics a population is a group of interbreeding individuals of the same species, which is isolated from other groups. In population ecology a population is a group of individuals of the same species inhabiting the same area (Sharov 1996).

✓   **Important Point**: Main axiom of population ecology—organisms in a population are ecologically equivalent. Ecological equivalency means:

1. Organisms undergo the same life cycle.
2. Organisms in a particular stage of the life cycle are involved in the same set of ecological processes.
3. The rates of these processes (or the probabilities of ecological events) are basically the same if organisms are put into the same environment (however some individual variation may be allowed).

When measuring populations, the level of species or density must be determined. Density (D) can be calculated by counting the number of individuals in the population (N) and dividing this number by the total units of space (S) the counted population occupies. Thus, the formula for calculating density becomes:

$$D = N/S$$

When studying aquatic populations, the occupied space (S) is determined by using length, width, and depth measurements. The volumetric space is then measured in cubic units.

Population density may change dramatically. For example, if a dam is closed off in a river midway through spawning season, with no provision allowed for fish movement upstream (a fish ladder), it would drastically decrease the density of spawning salmon upstream. Along with the swift and sometimes unpredictable consequences of change, it can be difficult to draw exact boundaries between various populations. Pianka (1988) makes this point in his comparison of European starlings that were introduced into Australia with starlings that were introduced into North America. He points out that these starlings are no longer exchanging genes with each other; thus, they are separate and distinct populations.

The population density or level of a species depends on natality, mortality, immigration, and emigration. Changes in population density are the result of both births and deaths. The birth rate of a population is called natality and the death rate, mortality. In aquatic populations, 2 factors besides natality and mortality can affect density. For example, in a run of returning salmon to their spawning grounds, the density could vary as more salmon migrate in, or as others leave the run for their own spawning grounds. The arrival of new salmon to a population from other places is termed *immigration (ingress)*. The departure of salmon from a population is called *emigration (egress)*. Thus, natality and immigration increase population density, whereas mortality and emigration decrease it. The net increase in population is the difference between these 2 sets of factors.

DISTRIBUTION (DISPERSION)

In ecology, **distribution** relates to where organisms are found on Earth—determined by biotic and/or abiotic factors. For example, large trees (pines, oaks, hickories) are not found in dry habitats like grasslands and deserts because there is not enough rainfall to sustain their growth—evolution has adapted trees to moist habitats. Moreover, organisms may be absent from a habitat due to the presence of predator species or competing species.

Where a particular species of organism does occur, the spatial relationships of individual organisms to one another may take several different forms—this is called **dispersion**. Dispersion is defined as the spatial distribution of individuals of a population. Each organism occupies only those areas that can provide for its requirements, resulting in an irregular dispersion. How a particular population is dispersed within a given area has considerable influence on density. As shown below, organisms in nature may be dispersed in 3 ways.

1. In **random dispersion**, there is an equal probability of an organism occupying any point in space, and "each individual is independent of the others" (Smith 1974). A randomly dispersed habitat is relatively uniform so individuals are neither repelled nor attracted to one another.

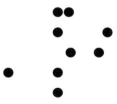

2. In a **regular** or **uniform distribution**, in turn, organisms are spaced more evenly; they are not distributed by chance; they are all about the same distance from one another. Animals compete with each other and effectively defend a specific territory, excluding other individuals of the same species. Another example is seen in desert shrubs that compete for water and often display a regular dispersion pattern. In regular or uniform distribution, the competition between individuals can be quite severe and antagonistic to the point where spacing generated is quite even (Odum 1983).

3. The most common distribution is the **contagious**, **clumped**, or **aggregated dispersion** where organisms are found in groups; this may reflect the heterogeneity of the habitat. Smith points out that contagious or clumped dispersions "produce aggregations, the result of response by plants and animals to habitat differences" (Smith 1974). Organisms that exhibit a contagious, clumped, or aggregated dispersion may develop social hierarchies in order to live together more effectively. Resources and suitable habitat may be patchy in dispersion, which causes organisms to form clumps (e.g., pillbugs living under rocks, or seals, who come together in clumps for breeding purposes).

✓   **Important Point**: Animals within the same species have evolved many symbolic aggressive displays that carry meanings that are not only mutually understood but also prevent injury or death within the same species. For example, in some mountainous regions, dominant male bighorn sheep force the juvenile and subordinate males out of the territory during breeding season. In this way, the dominant male gains control over the females and need not compete with other males.

POPULATION GROWTH

The size of animal populations is constantly changing due to natality, mortality, emigration, and immigration. As mentioned, the population size will increase if the natality and immigration rates are high. On the other hand, it will decrease if the mortality and emigration rates are high. Each population has an upper limit on size, often called the "carrying capacity." Carrying capacity can be defined as being the "optimum number of species' individuals that can survive in a specific area over time" (Enger et al. 1989). Stated differently, the carrying capacity is the maximum number of species that can be supported in a bioregion. A pond may be able to support only a dozen frogs depending on the food resources for the frogs in the pond. If there were 30 frogs in the same pond, at least half of them would probably die because the pond environment wouldn't have enough food for them to live. Carrying capacity is based on the quantity of food supplies, the physical space available, the degree of predation, and several other environmental factors.

The carrying capacity is of 2 types: ultimate and environmental. Ultimate carrying capacity is the theoretical maximum density; that is, it is the maximum number of individuals of a species in a place that can support itself without rendering the place uninhabitable. The environmental carrying capacity is the actual maximum population density that a species maintains in an area. Ultimate carrying capacity is always higher than environmental.

✓ **Important Point**: The population growth for a certain species may exhibit several types of growth.

Density-dependent factors are those which increase in importance as the size of the population increases. For example, as the size of a population grows, food and space may become limited. The population has reached the carrying capacity. When food and space become limited, growth is suppressed by competition. Density-dependent factors act "like governors on an engine and for this reason are considered one of the chief agents in preventing overpopulation" (Odum 1983).

✓ **Important Point**: Density-independent factors are those that have the same affect on population regardless of size. Typical examples of density-independent factors are devastating forest fires, streambeds drying up, or the destruction of the organism's entire food supply by disease. Thus, population growth is influenced by multiple factors. Some of these factors are generated within the population, others from without. Even so, usually no single factor can account fully for the curbing of growth in a given population. It should be noted, however, that humans are, by far, the most important factor; their activities can increase or exterminate whole populations.

POPULATION RESPONSE TO STRESS

As mentioned earlier, population growth is influenced by multiple factors. When a population reaches its apex of growth (its carrying capacity), certain forces work to maintain the population at a certain level. On the other hand, populations are exposed to small or moderate environmental stresses. These stresses work to affect the stability or persistence of the population. Ecologists have concluded that a major factor that affects population stability or persistence is species diversity. Species diversity is a measure of the number of species and their relative abundance.

Species diversity is related to several important ecological principles. For example, under normal conditions, high species diversity, with a large variety of different species, tends to spread risk. This is to say that ecosystems that are in a fairly constant or stable environment, such as a tropical rainforest, usually having higher species diversity. However, as Odum (1983) points out, "diversity tends to be reduced in stressed biotic communities."

If the stress on an ecosystem is small, the ecosystem can usually adapt quite easily. Moreover, even when severe stress occurs, ecosystems have a way of adapting. Severe environmental change to an ecosystem can result from such natural occurrences as fires, earthquakes, floods. and from people-induced changes such as land clearing, surface mining, and pollution.

Ecosystems can and do change. For example, if a forest is devastated by a fire, it will grow back, eventually, because of ecological succession. **Ecological succession** is the observed process of change (a normal occurrence in nature) in the species structure of an ecological community over time; that is, a gradual and orderly replacement of plant and animal species takes place in a particular area over time. The result of succession is evident in many places. For example, succession can be seen in an abandoned pasture. It can be seen in any lake and any pond. Succession can even be seen where weeds and grasses grow in the cracks in a tarmac, roadway, or sidewalk.

Below are additional specific examples of observable succession:

1. Consider a red pine planting area where the growth of hardwood trees (including ash, popular and oak) occurs. The consequence of this hardwood tree growth is the increased shading and subsequent mortality of the sun-loving red pines by the shade tolerant hardwood seedlings. The shaded forest floor conditions generated by the pines prohibit the growth of sun-loving pine seedlings and allow the growth of the hardwoods. The consequence of the growth of the hardwoods is the decline and senescence of the pine forest.
2. Consider raspberry thickets growing in the sunlit forest sections, beneath the gaps in the canopy generated by wind-thrown trees. Raspberry plants require sunlight to grow and thrive. Beneath the dense shade canopy—particularly of red pines but also dense stands of oak—there is not sufficient sunlight for the raspberry's survival. However, in any place in which there has been a tree fall, the raspberry canes proliferate into dense thickets. Within these raspberry thickets, by the way, are dense growths of hardwood seedlings. The raspberry plants generate a protected "nursery" for these seedlings and prevent a major browser of tree seedlings (the whitetail deer) from eating and destroying the trees. By providing these trees a shaded haven in which to grow, the raspberry plants are setting up the future tree canopy which will extensively shade the future forest floor and consequently prevent the future growth of more raspberry plants!

Succession usually occurs in an orderly, predictable manner. It involves the entire system. The science of ecology has developed to such a point that ecologists are now able to predict several years in advance what will occur in a given ecosystem. For example, scientists know that if a burned-out forest region receives light, water, nutrients, and an influx or immigration of animals and seeds, it will eventually develop into another forest through a sequence of steps or stages.

Two types of ecological succession are recognized by ecologists: primary and secondary. The particular type that takes place depends on the conditions at a particular site at the beginning of the process.

Primary succession, sometimes called bare-rock succession, occurs on surfaces such as hardened volcanic lava, bare rock, and sand dunes, where no soil exists, and where nothing has ever grown before. Obviously, in order to grow, plants need soil. Thus, soil must form on the bare rock before succession can begin. Usually this soil formation process results from weathering. Atmospheric exposure—weathering, wind, rain, and frost—forms tiny cracks and holes in rock surfaces. Water collects in the rock fissures and slowly dissolves the minerals out of the rock's surface. A pioneer soil layer is formed from the dissolved minerals and supports such plants as lichens. Lichens gradually cover the rock surface and secrete carbonic acid, which dissolves additional minerals from the rock. Eventually, the lichens are replaced by mosses. Organisms called decomposers move in and feed on dead lichen and moss. A few small animals such as mites and spiders arrive next. The result is what is known as a pioneer community. The pioneer community is defined as the first successful integration of plants, animals, and decomposers into a bare-rock community.

After several years, the pioneer community builds up enough organic matter in its soil to be able to support rooted plants like herbs and shrubs. Eventually, the pioneer community is crowded out and is replaced by a different environment. This, in turn, works to thicken the upper soil layers. The progression continues through several other stages until a mature or climax ecosystem is developed, several decades later. It is interesting to note that in bare-rock succession, each stage in the complex succession pattern dooms the stage that existed before it. According to Tomera (1989), "mosses provide a habitat most inhospitable to lichens, the herbs will eventually destroy the moss community, and so on until the climax stage is reached."

Secondary succession is the most common type of succession. Secondary succession occurs in an area where the natural vegetation has been removed or destroyed but the soil is not destroyed. For example, succession that occurs in abandoned farm fields, known as old field succession, illustrates secondary succession. Such an example can be seen in the Piedmont region of North Carolina. Early settlers of the area cleared away the native oak-hickory forests and cultivated the land. In the ensuing years, the soil became depleted of nutrients, reducing the soil's fertility. As a result, farming ceased in the region a few generations later, and the fields were abandoned. Some 150 to 200 years after abandonment, the climax oak-hickory forest was restored.

Flow of Energy in an Ecosystem

Simply defined, energy is the ability or capacity to do work. For an ecosystem to exist, it must have energy. All activities of living organisms involve work, which is the expenditure of energy. This means the degradation of a higher state of energy to a lower state. The flow of energy through an ecosystem is governed by 2 laws: the first and second laws of thermodynamics.

The first law, sometimes called the conservation law, states that energy may not be created or destroyed. The second law states that no energy transformation is 100% efficient. That is, in every energy transformation, some energy is dissipated as heat. The term *entropy* is used as a measure of the nonavailability of energy to a system. Entropy increases with an increase in dissipation. Because of entropy, the input of energy in any system is higher than the output or work done; thus, the resultant efficiency is less than 100%.

Odum (1983) explains that "the interaction of energy and materials in the ecosystem is of primary concern of ecologists." In biogeochemical nutrient cycles, it is the flow of energy that drives these cycles. Moreover, it should be noted that energy does not cycle as nutrients do in biogeochemical cycles. For example, when food passes from one organism to another, energy contained in the food is reduced step by step until all the energy in the system is dissipated as heat. Price (1984) refers to this process as "a unidirectional flow of energy through the system, with no possibility for recycling of energy." When water or nutrients are recycled, energy is required. The energy expended in this recycling is not recyclable. And, as Odum (1975) points out, this is a "fact not understood by those who think that artificial recycling of man's resources is somehow an instant and free solution to shortages."

As pointed out earlier, the principal source of energy for any ecosystem is sunlight. Green plants, through the process of photosynthesis, transform the sun's energy into carbohydrates which are consumed by animals. This transfer of energy, as stated previously, is unidirectional—from producers to consumers. Often this transfer of energy to different organisms is called a food chain.

All organisms, alive or dead, are potential sources of food for other organisms. All organisms that share the same general type of food in a food chain are said to be at the same trophic level (nourishment or feeding level). Since green plants use sunlight to produce food for animals, they are called the producers, or the 1st trophic level. The herbivores, which eat plants directly, are called the 2nd trophic level or the primary consumers. The carnivores are flesh eating consumers; they include several trophic levels from the 3rd on up. At each transfer, a large amount of energy (about 80% to 90%) is lost as heat and wastes. Thus, nature normally limits food chains to 4 or 5 links. However, in aquatic ecosystems, "food chains are commonly longer than those on land" (Dasmann 1984). The aquatic food chain is longer because several predatory fish may be feeding on the plant consumers. Even so, the built-in inefficiency of the energy transfer process prevents development of extremely long food chains.

Only a few simple food chains are found in nature. Most simple food chains are interlocked. This interlocking of food chains form a food web. A food web can be characterized as a map that shows what eats what (Miller 1988). Most ecosystems support a complex food web. A food web involves animals that do not feed on one trophic level. For example, humans feed on both plants and animals. The point is that an organism in a food web may occupy one or more trophic levels. Trophic level is determined by an organism's role in its particular community, not by its species. Food chains and webs help to explain how energy moves through an ecosystem.

An important trophic level of the food web that has not been discussed thus far comprises the decomposers. The decomposers feed on dead plants or animals and play an important role in recycling nutrients in the ecosystem. As Miller (1988) points out, "there is no waste in ecosystems. All organisms, dead or alive, are potential sources of food for other organisms."

FOOD CHAIN EFFICIENCY

Earlier, we pointed out that energy from the sun is captured (via photosynthesis) by green plants and used to make food. Most of this energy is used to carry on the plant's life activities. The rest of the energy is passed on as food to the next level of the food chain.

✓   **Important Point**: A food chain is the path of food from a given final consumer back to a producer.

It is important to note that nature limits the amount of energy that is accessible to organisms within each food chain. Not all food energy is transferred from one trophic level to the next. For ease of calculation, ecologists often assume an ecological efficiency of 10% to estimate the amount of energy transferred through a food chain. For example, if we apply the 10% rule to the diatoms-copepods-minnows-medium fish-large fish food chain, we can predict that 1,000 grams of diatoms produce 100 grams of copepods, which will produce 10 grams of minnows, which will produce 1 gram of medium fish, which, in turn, will produce 0.1 gram of large fish. Thus, only about 10% of the chemical energy available at each trophic level is transferred and stored in usable form at the next level. What happens to the other 90%? The other 90% is lost to the environment as low-quality heat in accordance with the second law of thermodynamics.

✓   **Important Point**: When an organism loses heat, it represents a one-way flow of energy out of the ecosystem. Plants only absorb a small part of energy from the sun. Plants store half of the energy and lose the other half. The energy plants lose is metabolic heat. Energy from a primary source will flow in one direction through 2 different types of food chains. In a grazing food chain, the energy will flow from plants (producers) to herbivores, and then through some carnivores. In detritus-based food chains, energy will flow from plants through detrivores and decomposers. In terms of the weight (or biomass) of animals in many ecosystems, more of their body mass can be traced back to detritus than to living producers. Most of the time the 2 food webs will intersect one another.

ECOLOGICAL PYRAMIDS

As we proceed in the food chain from the producer to the final consumer it becomes clear that a particular community in nature often consists of several small organisms associated with a smaller and smaller number of larger organisms. A grassy field, for example, has a larger number of grasses and other small plants, a smaller number of herbivores like rabbits, and an even smaller number of carnivores like foxes. The practical significance of this is that we must have several more producers than consumers.

This pound-for-pound relationship, where it takes more producers than consumers, can be demonstrated graphically by building an ecological pyramid. In an ecological pyramid, the number of organisms at various trophic levels in a food chain is represented by separate levels or bars placed one above the other, with a base formed by producers and the apex formed by the final consumer. The pyramid shape graphically represents the significant energy loss at each trophic level. The same is true if numbers of organisms are replaced by the corresponding biomass or energy. Ecologists generally use 3 types of ecological pyramids: pyramids of number, biomass, and energy. Obviously, there will be differences among them. Some generalizations:

1. **Energy pyramids** must always be larger at the base than at the top (based on the second law of thermodynamics, this pyramid represents the dissipation of energy as it moves from one trophic level to another).
2. Likewise, **biomass pyramids** (in which biomass is used as an indicator of production) are usually largest at the base. This is particularly true of terrestrial and aquatic systems dominated by large plants (marshes), in which consumption by heterotrophs is low and organic matter accumulates with time. It is important to point out, however, that biomass pyramids can sometimes be inverted. This is especially common in aquatic ecosystems, in which the primary producers are microscopic planktonic organisms that multiply very rapidly, have very short life spans, and are subject to heavy grazing by herbivores. At any single point in time, the amount of biomass in primary producers is less than that in larger, long-lived animals that consume primary producers.
3. **Numbers pyramids** can have various shapes (and may not be pyramids at all, actually) depending on the sizes of the organisms that make up the trophic levels. In forests, the primary producers are large trees and the herbivore level usually consists of insects, so the base of the pyramid is smaller than the herbivore level above it. In grasslands, the number of primary producers (grasses) is much larger than that of the herbivores above (large grazing animals).

To get a better idea of how an ecological pyramid looks and how it provides its information, we need to look at an example. The example to be used here is the energy pyramid. According to Odum (1983), the energy pyramid is a fitting example because among the "three types of ecological pyramids, the energy pyramid gives by far the best overall picture of the functional nature of communities."

In an experiment conducted in Silver Springs, Florida, Odum measured the energy for each trophic level in terms of kilocalories. A kilocalorie is the amount of energy needed to raise 1 cubic centimeter of water 1 degree centigrade. When an energy pyramid is constructed to show Odum's findings, it takes on the typical upright form (as it must because of the second law of thermodynamics).

Simply put, according to the second law of thermodynamics, no energy transformation process is 100% efficient. This fact is demonstrated, for example, when a horse eats hay. The horse cannot obtain, for his own body, 100% of the energy available in the hay. For this reason, the energy productivity of the producers must be greater than the energy production of the primary consumers. However, if human beings are substituted for the horse, and the horse is substituted for the hay, it is interesting to note that according to the second law of thermodynamics, only a small population could be supported. But this is not the case. Humans also feed on plant matter, which allows a larger population. Therefore, if meat supplies become scarce, we must eat more plant matter. This is the situation we see today in countries where meat is scarce. Consider this, if we all ate soybean, there would be at least enough food for 10 times as many of us as compared to a world where we all eat beef (or pork, fish, chicken, etc.).

PRODUCTIVITY

As mentioned previously, the flow of energy through an ecosystem starts with the fixation of sunlight by plants through photosynthesis. In evaluating an ecosystem, the measurement of photosynthesis is important. Ecosystems may be classified into highly productive or less productive. Therefore, the study of ecosystems must involve some measure of the productivity of that ecosystem.

Smith defines production (more specifically, primary production, because it is the basic form of energy storage in an ecosystem) as being "the energy accumulated by plants" (Smith 1974). Stated differently, primary production is the rate at which the ecosystem's primary producers capture and store a given amount of energy, in a specified time interval. In even simpler terms, primary productivity is a measure of the rate at which photosynthesis occurs. Odum (1971) lists 4 successive steps in the production process as follows:

1. **Gross primary productivity**—the total rate of photosynthesis in an ecosystem during a specified interval.
2. **Net primary productivity**—the rate of energy storage in plant tissues in excess of the rate of aerobic respiration by primary producers.
3. **Net community productivity**—the rate of storage of organic matter not used.
4. **Secondary productivity**—the rate of energy storage at consumer levels.

When attempting to comprehend the significance of the term "productivity" as it relates to ecosystems, it is wise to consider an example. Consider the productivity of an agricultural ecosystem such as a wheat field. Often its productivity is expressed as the number of bushels produced per acre. This is an example of the harvest method for measuring productivity. For a natural ecosystem, several 1 square meter plots are marked off, and the entire area is harvested and weighed to give an estimate of productivity as grams of biomass per square meter per given time interval. From this method, a measure of net primary production (net yield) can be measured.

Productivity, both in the natural and cultured ecosystem, may vary considerably, not only between type of ecosystems, but also within the same ecosystem. Several factors influence year-to-year productivity within an ecosystem. Such factors as temperature, availability of nutrients, fire, animal grazing, and human cultivation activities are directly or indirectly related to the productivity of a particular ecosystem.

The ecosystem that is of greatest importance in this particular study is the aquatic ecosystem. Productivity can be measured in several different ways in the aquatic ecosystem. For example, the production of oxygen may be used to determine productivity. Oxygen content may be measured in several ways. One way is to measure it in the water every few hours for a period of 24 hours. During daylight, when photosynthesis is occurring, the oxygen concentration should rise. At night the oxygen level should drop. The oxygen level can be measured by using a simple x-y graph. The oxygen level can be plotted on the y-axis with time plotted on the x-axis.

Another method of measuring oxygen production in aquatic ecosystems is to use light and dark bottles. Biochemical oxygen demand (BOD) bottles (300 ml) are filled with water to a particular height. One of the bottles is tested for the initial dissolved oxygen (DO), and then the other two bottles (one clear, one dark) are suspended in the water at the depth they were taken from. After a 12-hour period, the bottles are collected and the DO values for each bottle are recorded. Once the oxygen production is known, the productivity in terms of grams/m/day can be calculated.

THE BOTTOM LINE

The ecological trends paint a clear picture. Wherever we look, ecological productivity is limping behind human consumption. Since 1984, the global fish harvest has been dropping, and so has the per capita yield of grain crops (Brown 1994). Moreover, stratospheric ozone is being depleted; the release of greenhouse gases has changed the atmospheric chemistry and might lead to climate change; erosion and desertification is reducing nature's biological productivity; irrigation water tables are falling; contamination of soil and water is jeopardizing the quality of food; other natural resources are being consumed faster than they can regenerate; and biological diversity is being lost—to reiterate only a small part of a long list. These trends indicate a decline in the quantity and productivity of nature's assets (Wackernagel 1997).

Chapter Review Questions

1. A web of complex _____ binds all living things in any region.
2. _____ are large-scale communities classified by a main vegetation type and distinctive combinations of plants and animals.
3. _____ cycles are driven by energy, directly or indirectly from the sun.
4. A biome is a larger region than _____.
5. Area of less than 25m/yr rainfall: _____.
6. Land of short and tall grass but no trees: _____.
7. Boreal forest: _____.
8. _____ relates to where organisms are found on Earth.
9. _____ is defined as being the optimum number of a species' individuals that can survive in a specific area over time.
10. _____ is the observed process of change in the species structure of an ecological community over time.

References and Recommended Reading

Abrahamson, D.E., ed. 1988. *The Challenge of Global Warming*. Washington, DC: Island Press.

Adams, V.D. 1990. *Water and Wastewater Examination Manual*. Chelsea, MI: Lewis Publishers, Inc.

Allen, J.D. 1996. *Stream Ecology: Structure and Function of Running Waters*. London: Chapman and Hall.

Barbour, M.T., J. Gerritsen, B.D. Snyder, and J.B. Stribling. 1997. *Revision to Rapid Bioassessment Protocols for Use in Streams and Rivers, Periphyton, Benthic Macroinvertebrates and Fish*. Washington, DC: United States Environmental Protection Agency. EPA 841-D-97-002, 1997.

Berryman, A.A. 1981. *Population Systems: A General Introduction*. New York: Plenum Press.

Bradbury, I. 1991. *The Biosphere*. New York: Belhaven Press.

Brown, L.R. 1994. Facing Food Insecurity. In *State of the World*. New York: W.W. Norton, pp. 179–187.

Camann, M. 1996. *Freshwater Aquatic Invertebrates: Biomonitoring*. Accessed July 7, 2006, at www.humboldt.edu.

Carson, R. 1962. *Silent Spring*. Boston: Houghton Mifflin Company.

Clark, L.R., P.W. Gerier, R.D. Hughes, and R.F. Harris.1967. *The Ecology of Insect Populations*. London: Methuen.

Cummins, K.W. 1994. Structure and Function of Stream Ecosystems. *Bioscience* 24: 631–641.

Cummins, K.W., and M.J. Klug.1979. Feeding Ecology of Stream Invertebrates. *Annual Review of Ecology and Systematics* 10: 631–641.

Darwin, C. 1998. *The Origin of Species*, ed. Suriano, G. New York: Grammercy.

Dasmann, R.F. 1984. *Environmental Conservation*. New York: John Wiley and Sons.

Davis, M.L., and D.A. Cornwell. 1991. *Introduction to Environmental Engineering*. New York: McGraw-Hill Inc.

Enger, E., J.R. Kormelink, B.F. Smith, and R.J. Smith. 1989. *Environmental Science: The Study of Interrelationships*. Dubuque, IA: Wm. C. Brown Publishers.

Evans, E.D., and H.H. Neunzig. 1996. Megaloptera and Aquatic Neuroptera. In *Aquatic Insects of North America*, eds. Merritt, R.W., and K.W. Cummins, 3rd ed. Dubuque, IA: Kendall/Hunt Publishing Company, pp. 298–308.

Evans, R. 1965. Industrial Wastes and Water Supplies. *Journal American Water Works Association* 57: 625–628.

Freedman, B. 1989. *Environmental Ecology*. New York: Academic Press, Inc.

Hamburg, M. 1987. *Statistical Analysis for Decision Making*. New York: Harcourt Brace Jovanovich, Publishers.

Hewitt, C.N., and R. Allott. 1992. *Understanding Our Environment: An Introduction to Environmental Chemistry and Pollution*. Cambridge, UK: The Royal Society of Chemistry.

Hickman, C.P., L.S. Roberts, and F.M. Hickman. 1990. *Biology of Animals*. St. Louis, MO: Times Mirror/Mosby College Publishing.

Hickman, C.P., L.S. Roberts, and F.M. Hickman. 1998. *Integrated Principles of Zoology*. St. Louis, MO: Times Mirror/Mosby College Publishing.

Jeffries, M., and D. Mills. 1990. *Freshwater Ecology: Principles and Applications*. London: Belhaven Press.

Karr, C. 1991. Biological Integrity and the Goal of Environmental Legislations: Lessons for Conservation Biology. *Conservation Biology* 4:66–84.

Karr, J.R., and D.R. Dudley. 1981. Ecological Perspective on Water Quality Goals. *Environmental Management* 5: 55–68.

Kimmel, W.G. 1983. The Impact of Acid Mine Drainage on the Stream Ecosystem. In *Pennsylvania Coal: Resources, Technology, and Utilization*. eds. S. K. Majumdar and W.W. Miller. The Pennsylvania Academy of Science, pp. 424–437.

Lafferty, P., and J. Rowe. 1993. *The Dictionary of Science*. New York: Simon & Schuster.

Leopold, L.B. 1970. *A View of the River*. Cambridge, MA: Harvard University Press.

Mackie, G.I. 1998. *Applied Aquatic Ecosystem Concepts*. Quelph, Canada: University of Guelph.

Madsen, J. 1985. *Up on the River*. New York: Lyons Press.

Mason, C.F. 1990. *Biological Aspects of Freshwater Pollution. Pollution: Causes, Effects, and Control*. Cambridge, UK: The Royal Society of Chemistry.

Masters, G.M. 1991. *Introduction to Environmental Engineering and Science*. Englewood Cliffs, NJ: Prentice Hall.

McCafferty, P.W. 1981. *Aquatic Entomology, the Fishermen's and Ecologists' Illustrated Guide to Insects and Their Relatives*. Boston: Jones and Bartlett Publishers.

McGhee, T.J. 1991. *Water Supply and Sewerage*. New York: McGraw-Hill, Inc.

Michaud, J.P. 1994. *A Citizen's Guide to Understanding and Monitoring Lakes and Streams. Publications #94–149*. Olympia: Washington State Department of Biology, pp. 1–13.

Miller, G.T. 1988. *Environmental Science: An Introduction*. Belmont, CA: Wadsworth Publishing Company.

Moran, J.M., M.D. Morgan, and J.H. Wiersma.1986. *Introduction to Environmental Science*. New York: W.H. Freeman and Company.

Naar, J. 1990. *Design for a Livable Planet*. New York: Harper and Row, Publishers.

O'Connor, D., and W. Dobbins. 1975. The Mechanism of Reaeration in Natural Streams. *Journal of Hydraulics Division, Proceedings of American Society of Civil Engineers* 101:1315.

Odum, E.P. 1971. *Fundamentals of Ecology*. Philadelphia: Saunders College Publishing.

Odum, E.P. 1975. *Ecology: The Link Between the Natural and the Social Sciences*. New York: Holt, Rinehart and Winston, Inc.

Odum, E.P. 1983. *Basic Ecology*. Philadelphia: Saunders College Publishing.

Overcash, M.R., and J.M. Davidson. 1981. *Environmental Impact of Nonpoint Source Pollution*. Ann Arbor, MI: Ann Arbor Science Publishers, Inc.

Pennack, R.W. 1989. *Fresh-Water Invertebrates of the United States*, 3rd ed. New York: John Wiley & Sons, Inc.

Pianka, E.R. 1988. *Evolutionary Ecology*. New York: Harper Collins Publisher.

Porteous, A. 1992. *Dictionary of Environmental Science and Technology*. New York: John Wiley and Sons, Inc.

Price, P.W. 1984. *Insect Ecology*. New York: John Wiley and Sons.

Sharov, A. 1996. *What Is Population Ecology?* Blacksburg, VA: Department of Entomology, Virginia Tech University.

Sharov, A. 1997. *Population Ecology*. Blacksburg, VA: Department of Entomology, Virginia Tech University. Lecture notes accessible at www.gypsymoth.ento.vt.edu/~sharov/popechome/welcome.html

Sharov, A.A. 1992. Life-System Approach: A System Paradigm in Population Ecology. *Oikos* 63: 485–494.

Smith, R.A., R.B. Alexander, R.B., and M.G. Wolman. 1987. Water-Quality Trends in the Nation's Rivers. *Science* 235:1607–1615.

Smith, R.L. 1974. *Ecology and Field Biology*. New York: Harper and Row, Publishers.

Stewart, K.W., and B.P Stark. 1998. *Nymphs of North American Stonefly General (Plecoptera)*, Volume 12. Denton, TX: Thomas Say Foundation.

Tansley, A.G. 1935. The Use and Abuse of Vegetational Concepts and Terms. *Ecology* 16, 284–307.

Tomera, A.N. 1989.*Understanding Basic Ecological Concepts*. Portland, ME: J. Weston Walch, Publisher.

Wackernagel, M. 1997. *Framing the Sustainability Crisis: Getting from Concerns to Action*. www.sdri.ubc.ca/publications/wacherna.html.

Welch, P.S. 1963. *Limnology*. New York: McGraw-Hill Book Company.

Westman, W.E. 1985. *Ecology, Impact Assessment, and Environmental Planning*. New York: John Wiley and Sons, Inc.

Wetzel, R.G. 1983. *Limnology*. New York: Harcourt Brace Jovanovich College Publishers.

Wooton, A. 1984. *Insects of the World*. New York: Facts on File, Inc.

WRI and IIED. 1986. *World Resources 1986*. New York: World Resource Institute and International Institute of Environment and Development.

WRI and IIED. 1988. *World Resources 1988–1989*. New York: World Resource Institute and International Institute of Environment and Development.

Answer Key for
Chapter Review Questions

Chapter 1

1. The study of life.
2. Living things: are composed of cells; reproduce; respond to stimuli; maintain homeostasis; require energy; display heredity; evolve and adapt.
3. Amino acids.
4. Enzymes.
5. Nucleic acids.
6. Atoms, molecules, organelles, cell, tissue, organ, organ system, organism.
7. Species, population, community, ecosystem, biome, biosphere.
8. All life-forms are made from one or more cells; cells only arise from pre-existing cells; the cell is the smallest form of life.
9. Initial observations and objectives; hypothesis formulation; data collection, analysis of data; summarization of results; discussion of limitations and conclusions; identification of future research needs.
10. The organism must be found in all animals suffering from the disease, but not in healthy animals; the organism must be isolated from a diseased animal and grown in pure culture; the cultured organism should cause disease when introduced into a healthy animal; the organisms must be isolated from the experimentally infected animal.

Chapter 2

1. Health care, cooking, cosmetics, automobiles.
2. Matter is anything that takes up space and has mass.
3. A substance that cannot be broken down or decomposed into simpler forms of matter.
4. Substances that can be broken down into 2 or more simpler substances.

5. Metals.
6. Metalloids.
7. A change in which the molecular structure of a substance is not altered.
8. Atomic weight.
9. A strong force of attraction holding atoms together in a molecule.
10. Salt.
11. Carbon.
12. Small characteristic groups of atoms that are frequently bonded to the carbon skeleton of organic molecules.
13. Carbohydrates, lipids, proteins, nucleic acids.
14. Carbohydrates.
15. Lipids.

Chapter 3

1. Proteins embedded in a phospholipid bilayer.
2. Made of the protein tubulin; are involved in providing motility; are organized by centrioles.
3. Degradation.
4. Mitochondria.
5. Cell walls.
6. Lipids.
7. Chitin, polysaccharides, cellulose, peptidoglycans.
8. Prokaryotes and eukaryotes.
9. Chloroplasts.

Chapter 4

1. A series of biochemical, enzyme-mediated reactions during which atmospheric carbon dioxide is reduced and incorporated into organic molecules.
2. One of the 3 phases of the catabolism of glucose to carbon and water process.
3. Process by which a compound is oxidized using oxygen as an external electron acceptor.
4. Energy cannot be created nor destroyed.
5. Heat flows from regions of high temperatures to regions of low temperatures.
6. Adenosine triphosphate.
7. Adenosine diphosphate.
8. Nicotinamide adenine dinucleotide.
9. Nonprotein groups which act as catalytic activators for certain enzymes
10. When the cofactor is loosely attached to the apoenzyme.
11. The energy wheel of cellular respiration.
12. A molecule synthesized in our bodies from the vitamin riboflavin.

Chapter 5

1. Mitosis and meiosis.
2. To the centromere via the kinetochore.
3. Interphase, G1, S, G2, and M phase (the last includes prophase, metaphase, anaphase, telophase).
4. Prophase, metaphase, anaphase, telophase.
5. Two.
6. Prophase I, metaphase I, anaphase I, telophase I, prophase II, metaphase II, anaphase II, telophase II.
7. Homologous.
8. Four.
9. Meiosis reduces the chromosome number in half to produce haploid cells (i.e., each daughter cell has half the number of chromosomes as the original cell).
10. A gene is a discrete unit of heredity located on chromosomes and a gene consists of DNA. An allele is an alternate form of a specific gene.
11. Genotype is the genetic composition of an individual, especially in terms of the alleles for particular genes. Phenotype is the outward appearance of an organism.
12. Dominant describes an allele that determines the genotype in the heterozygous condition. Recessive describes an allele that is masked by the presence of a dominant allele.

Chapter 6

1. Binomial system of nomenclature
2. Monera, Protista, Fungi, Plantae, Animalia
3. Protista
4. Fungi
5. Bacteria
6. Bacilli
7. Bacteriophage
8. Lyses
9. Protozoa
10. The pellicle
11. Algae
12. Algae
13. Frustules
14. Chrysophyta
15. Pyrrophyta
16. Budding
17. Diploid
18. Tracheophytes
19. Chloroplasts

20. Cambia
21. Phototropism
22. Photoperiodism
23. Invertebrates
24. Annelids
25. Invertebrate chordates

Chapter 7

1. Epithelial, connective, nervous, and muscle
2. Epithelial
3. Any 3 of the following: providing physical protection, controlling permeability absorption, and producing specialized secretions
4. Collagen
5. Bones
6. Histology
7. Muscle
8. Adipocytes
9. Serous

Chapter 8

1. Skin, hair, nails, epithelial glands
2. Blood
3. Papillary and reticular
4. Keratin
5. Nucleus
6. In the epidermis
7. Any 6 of the following: protection, body temperature regulation, cutaneaus sensation, metabolic functions, blood reservoirs, excretion, absorption route
8. Hair follicles
9. Keratinocytes and melanocytes
10. Melanin, carotene, and hemoglobin

Chapter 9

1. Blood
2. Plasma
3. Albumin
4. RBCs
5. WBCs

6. Monocytes
7. Blood clots
8. Arteries
9. Arterioles
10. Superior vena cava and inferior vena cava

Chapter 10

1. Lymph nodes
2. Lymphocytes
3. Lymphatic system
4. Helper T cells
5. Monocytes
6. Bone marrow
7. Stem cells
8. Antibodies
9. Phagocytes
10. Mast cell
11. Complement system
12. Clonal anergy

Chapter 11

1. Mouth
2. Small intestine
3. Stomach
4. Liver, pancreas
5. Liver
6. Digested fat
7. Cecum
8. Sigmoid colon
9. Saliva
10. Lactose
11. Gastrin
12. Peptide yy

Chapter 12

1. Central nervous system
2. Peripheral nervous system
3. Afferent, efferent

4. Interneuron (or association)
5. Sympathetic
6. Soma
7. Axons
8. Nissl substance
9. Oligodendrocytes
10. Multiple sclerosis, myelin sheath
11. Ependymal cells, cerebrospinal fluid

Chapter 13

1. Endocrine
2. Adrenal gland
3. Homeostasis
4. Pineal gland
5. Exocrine
6. Tropic hormones
7. Oxytocin
8. Thyroxine
9. Adrenal medulla
10. Glucagon
11. Gonads
12. Type 1

Chapter 14

1. Reproductive
2. Ovaries and testes
3. Only half
4. Spermiogenesis
5. Epididymis
6. Meiosis
7. Sweat glands
8. Amniotic sac
9. Ovum
10. Vulva
11. Ovary

Chapter 15

1. Muscle contraction
2. Such as posture, joint stability, heat production

3. True
4. Epimysium, perimysium, endomysium
5. Autonomic
6. Soft body parts
7. Calcium
8. Haversian canal
9. True
10. Spongy bone

Chapter 16

1. Internal
2. Atmosphere, gases
3. Age, sex, size
4. Nares
5. Pharynx
6. False
7. COPD
8. Bronchi
9. False
10. Pulmonary ventilation

Chapter 17

1. Excretory
2. True
3. Blood, wastes, excrete
4. Nephrons
5. Urine
6. Sphincters
7. Calyces
8. Sweat
9. Rennin
10. Urinary

Chapter 18

1. Adaptation
2. Adaptive radiation
3. Microevolution
4. Speciation

5. *Scala naturae*, also called the Great Chain of Being or Ladder of Nature
6. Linnaeus
7. Lamarck
8. Cuvier
9. Natural selection
10. Hardy-Weinberg Principle

Chapter 19

1. Relations
2. Biomes
3. Biogeochemical
4. An ecosystem
5. Desert
6. Savanna
7. Taiga
8. Distribution
9. Carrying capacity
10. Ecological succession

Index

About the Author

Frank R. Spellman is an assistant professor/lecturer of environmental health at Old Dominion University in Norfolk, Virginia. He has more than thirty-five years of experience in environmental science and engineering, in both the military and the civilian communities. A professional member of the American Society of Safety Engineers, the Water Environment Federation, and the Institute of Hazardous Materials Managers, Frank Spellman is a Board Certified Safety Professional (CSP) and Board Certified Hazardous Materials Manager (CHMM). He has authored/co-authored fifty-one texts on safety, occupational health, water/wastewater operations, environmental science, and concentrated animal feeding operations.